VOICE OVER IP NETWORKS

Marcus Goncalves

McGraw-Hill

New York San Francisco Washington, D.C.
Auckland Bogotá Caracas Lisbon London
Madrid Mexico City Milan Montreal New Delhi
San Juan Singapore Sydney Tokyo Toronto

McGraw-Hill

A Division of The **McGraw-Hill** *Companies*

Copyright © 1999 by The McGraw-Hill Companies, Inc. All rights reserved. Printed in the United States of America. Except as permitted under the United States Copyright Act of 1976, no part of this publication may be reproduced or distributed in any form or by any means, or stored in a data base or retrieval system, without the prior written permission of the publisher.

The views expressed in this book are solely those of the author, and do not represent the views of any other party or parties.

2 3 4 5 6 7 8 9 0 AGM/AGM 9 0 3 2 1 0 9

P/N 024829-X

Part of ISBN 0-07-913783-0

The sponsoring editor for this book was Simon Yates and the production supervisor was Tina Cameron. It was set in Galliard by Patricia Wallenburg.

Printed and bound by Quebecor/Martinsburg.

McGraw-Hill books are available at special quantity discounts to use as premiums and sales promotions, or for use in corporate training programs. For more information, please write to the Director of Special Sales, McGraw-Hill, 11 West 19th Street, New York, NY 10011. Or contact your local bookstore.

Information contained in this work has been obtained by The McGraw-Hill Companies, Inc. ("McGraw-Hill") from sources believed to be reliable. However, neither McGraw-Hill nor its authors guarantee the accuracy or completeness of any information published herein and neither McGraw-Hill nor its authors shall be responsible for any errors, omissions, or damages arising out of use of this information. This work is published with the understanding that McGraw-Hill and its authors are supplying information but are not attempting to render engineering or other professional services. If such services are required, the assistance of an appropriate professional should be sought.

This book is printed on recycled, acid-free paper containing a minimum of 50% recycled, de-inked fiber.

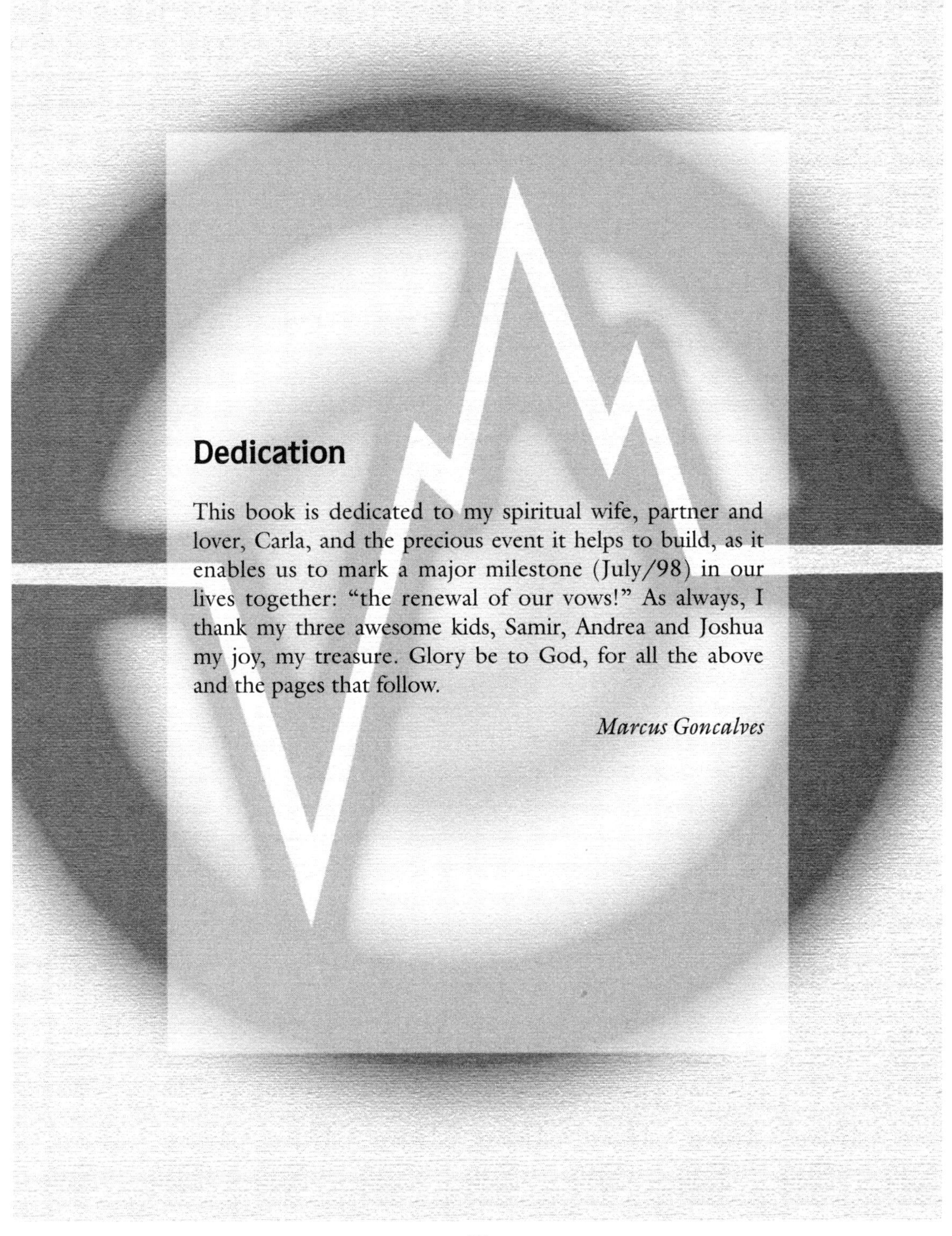

Dedication

This book is dedicated to my spiritual wife, partner and lover, Carla, and the precious event it helps to build, as it enables us to mark a major milestone (July/98) in our lives together: "the renewal of our vows!" As always, I thank my three awesome kids, Samir, Andrea and Joshua my joy, my treasure. Glory be to God, for all the above and the pages that follow.

Marcus Goncalves

Acknowledgments

There are many people I would like to acknowledge, as I'm very thankful to so many. Since I must start somewhere, I would like to thank those who went an extra mile in making this project a joyful one.

I thank Stephen Thomas, CTO of TransNexus, for kindly sharing with me information about his presentation on to the VoIP Forum; Terry Young, of Vienna Systems, for his great support in providing information about Vienna.way technology. I also thank Lisa Hampell and Tara Krull for their great support in providing me information on Qwestís Macro Capacity Fiber Network, a state-of-the-art infrastructure fully supporting VoIP applications.

My thanks to Andy Voss, of Nuera Communications, for all his support in providing me full access to Nueraís site and information about their many VoIP technologies and products. On the same note, I thank Stacy Doherty of Micom (a Nortel company), for her wiliness and support during my preliminary research and information about Nortel's Internet Voice Button technology, and the people at Lucent, NetSpeak and Vocaltec for their support to this project.

Many thanks goes to the authors of RFC 2326, Rob Lanphier of RealNetworks, Henning Schulzrinne of Columbia University, and Anup Rao of Netscape, for their support and technical expertise, allowing me to use the RFC to present and discuss RTSP and getting me interested in the subject in the first place.

Most of all, a special thank you to Phil Emer, from North Carolina State University, for his tremendous support on ATM/VoIP technology, as his impressive knowledge a direct contribution greatly added to the quality of the information in this book; and also George Petrolekas, of ABL, for his great support and kindness in making his paper on video over frame relay available to me, which also added to the depth of information in this book.

Lastly I thank my beautiful wife Carla, for her continued encouragement, support, and patience, and the people at McGraw-Hill, especially Simon Yates, for entrusting this project to me. Most importantly, I thank God for allowing me to contribute to a better (virtual) world this way.

Foreword

Watching television the other day, one of my favorite commercial came on air from a well known investment company discussing the pace of technology. The key line for me was "a seven year old airplane is younger than a one year old computer." The pace of change, of what is possible, is now measured in months not years.

As you read Marcus Goncalves' insightful book on Voice Over IP always remember that line. Marcus has done what many for centuries have tried to do through the use of the stars or cards or even tea leaves; he has given us a glimpse into the very near future. As technology professionals or even as common consumers we should take heed of what he says.

As a member of several technology forums, I have seen the writing on the wall. It is clear to me that the technological future lies with IP routing. There are many who argue that IP based applications are "not ready for prime time". To those who espouse that point of view, I can only say that you risk being left at the side of the road. A quick look at our world will confirm that.

In the 50's, television really appeared as a mass market communications medium. It brought information into the home. Originally, it was only available to a select few because of high cost, and the lack of a sufficiently large installed base, limited the number of channels and content which was available. Many at that time argued in favor of radio as television was not ready for prime time. Slowly, prices came down and more and more households had televisions. Color Television arrived and the same process was repeated, only now, households had at least two television sets. Today, there are hardly any homes regardless of economic grouping that do not have at least one television at home; and this applies around the world.

The exact same evolution can be seen with computers, especially with the release to market of sub $1,000 PC's and now $500 web PC's. The concurrent trend is the explosion of the web in terms of penetration and as a content provider/delivery system like television.

Enough future lies with IP routing as everything is becoming digital and hence data, non-data and other types of particularly as concerns the Internet. Therefore, as The Economist pointed out, telephone companies will have no choice but to join in.

Through Marcus' crystal ball, we too can look into the future. Marcus points out the applications made possible, but ties together not only IP telephony, but the various networks which exist in the present day such as Frame Relay, ATM and SONET backbones, and how Voice Over IP ties into all of these infrastructures. He also explains in detail the underlying technology and it's upcoming evolution as well as examining the equipment options available on which to base networking decisions.

Today as I wrote this, I went to my local telco to inquire about installation of residential ISDN for my new home. After a short talk with the telco sales rep, he informed me that I should wait till September when the telco would introduce xDSL services with multiples of bandwidth at less cost than the single BRI I was about to have installed. I can only imagine the multitude of services I will now be able to install now that I no longer have a bandwidth restriction. Only three months ago, the 56k modem was ratified by the ITU-T. Sometimes, the future comes faster than we think.

George Petrolekas
Director, Marketing
AB1 Canada Inc.
Montreal
June 1998

Table of Contents

Preface

Voice Over IP Networks thoroughly explores the potential and increasing demand of delivering Internet-based telephony as the next technological and profitable frontier for every company running a TCP/IP stack network. Up to now, the Internet has run over phone lines, but voice over IP technologies reverse this trend by providing telephony services over the Internet.

This book was developed with IPv6 in mind, exploring and explaining how these two cutting-edge technologies will merge to deliver powerful telephony-based services and applications. It does take IPv4 into consideration and discusses the complexity of deploying this technology, due to the connectionless nature of IP networks and the engineering challenges they engender.

This book also thoroughly explores the potential investment in terms of both money and technology. It provides in-depth discussion of the main practical implementation and solutions being advocated by leading vendors and their products, such as 3Com's VoIP implementation Total Control™ HiPer™ Access System, Motorola's VIPR, Nuera Communications' Access Plus F200ip Voice FRAD/VOIC and others.

By the end of this book the reader should have a full understanding of how VoIP (Voice Over Internet Protocol) works, the main challenges in implementing it and the major vendors and their products. The reader will also have a reasonable knowledge of the main products and implementation solutions being proposed.

Overview

The book is divided in three parts.

Part I, The Technology provides basic technical grounds for understanding IP and the voice transferring process over it.

Chapter 1, An Overview of IPv4 and IPv6, provides information about the IPv4's strengths and weaknesses, as well as coverage of IPv6 and its enhanced internetworking capabilities.

Chapter 2, Understanding RSVP, IP Multicast and ATMs, discusses the fundamentals of RSVP and IP Multicast and gives an overview of ATM; all these are important technologies for carrying voice over IP.

Chapter 3, IP Superhighway, introduces the basic concepts of voice over IP, and its most-used H.323 standard. It also discusses other standards and technologies such as audio codecs, IP over ATM, voice over ATM, the emulation of traditional T1/E1 Trunks, IP over SONET and voice over SONET, and IP and voice over frame relay. This chapter also treats Layer 3 switching and gigabit Ethernet as well as their role in VoIP.

Chapter 4, More on IP Multicasting, discusses multicasting in workgroups and some of its capabilities on hosts and routers, as well as usage and implementation, especially with VoIP.

Chapter 5, More on ATM Technologies, discusses the ATM data model, its network services, data protocols and LAN emulation, as well as ATM MPOA services.

Chapter 6, Broadband Packet Networks and Voice Communication, discusses broadband packet networks and their role in voice communication and applications.

Part II, Hands-on VoIP: Standards and Implementations, is a practical section discussing issues as well as proposed scenarios.

Chapter 7, Codecs Methods, covers the technology and standards employed in voice digitization and reviews video and audio codecs.

Chapter 8, Voice Over IP: Can We Talk?, focus on the applicability of VoIP, as in computer telephony integration (CTI), videoconferencing, document-sharing, Web-based call center applications, etc. It also discusses the challenges VoIP faces, in getting telcos up to speed, and in setting standards. It lists the major VoIP players, including 3Com, Motorola, Nuera Communications and others.

Chapter 9, What to Expect: The Innovators, assesses what is being offered by VoIP innovators such as NetSpeak, NetPhone, Vocaltec, TeleVideo Conversions, Inc., Vienna Systems, Lucent Technologies, and others.

Part III, Advanced VoIP, outlines the major players of the industry, their products, technical characteristics and specifications.

Chapter 10, The RTSP Protocol, assesses the Real Time Streaming Protocol (RTSP).

There are aso two appendices: Appendix A provides a list of VoIP vendors and Appendix B provides a comprehensive glossary of terms related to VoIP.

Who Should Read This Book

This book is designed for systems managers, network administrators, systems integrators, Internet managers and even chief information officers implementing or planning to implement VoIP in their businesses. It provides a brief general review of basic concepts of IPv4, IPv6 and VoIP, for those not too familiar with IP. It also provides advanced configuration and troubleshooting information for professionals heavily involved with VoIP projects as well as an outline of the main products available on the market.

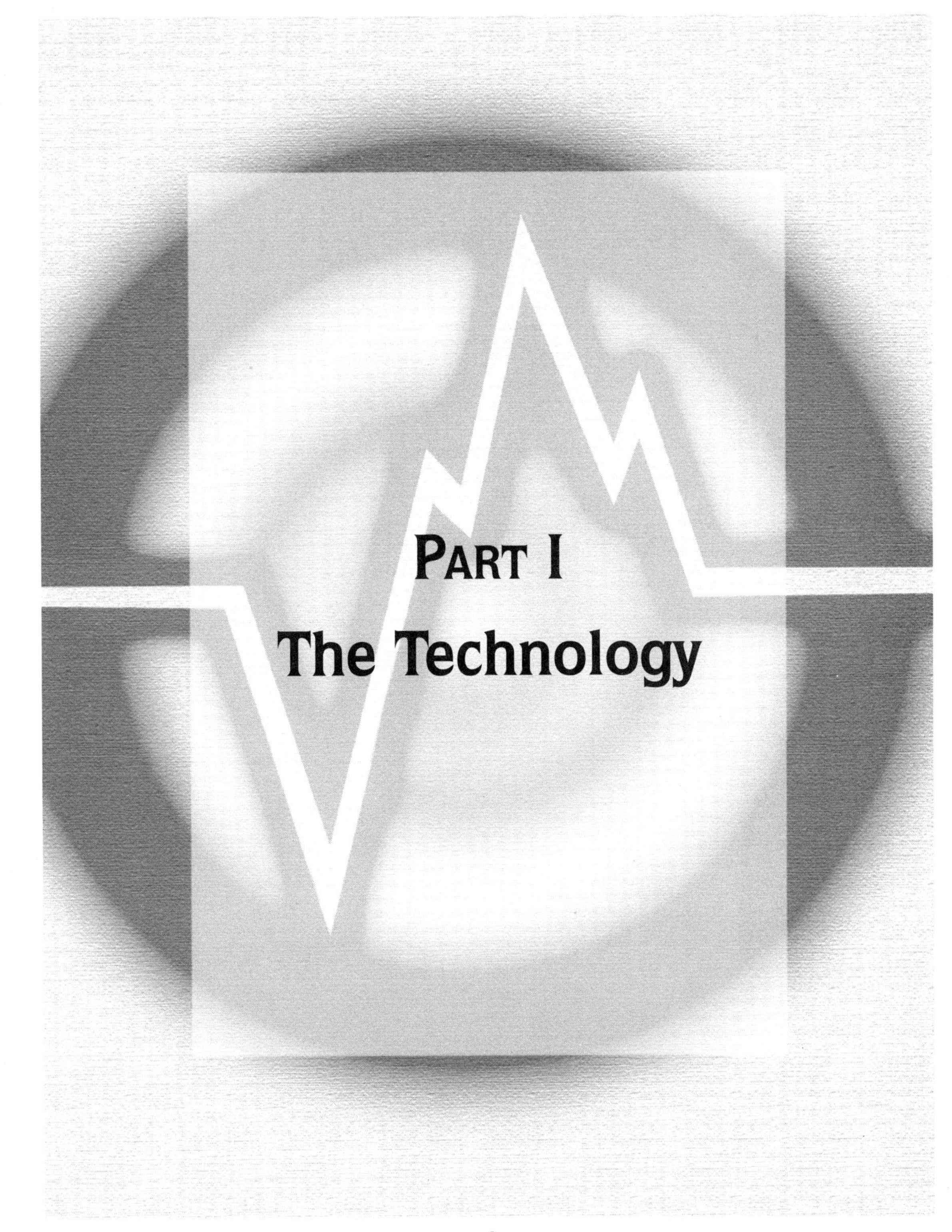

PART I
The Technology

CHAPTER 1

An Overview of IPv4 and IPv6

IPv4 was developed in 1975 and provides for approximately 4.2 billion possible combinations. Although this sounds like more than enough addresses, the truth is that every machine as well as every interface on every machine requires a unique address. We therefore find ourselves in our current position of lacking capacity in addresses to cover the impending supply of users.

By the beginning of the new century, the Internet community will need enough IP addresses for the billions upon billions of new customers that it attracts as well as the possible new hosts being setup and connected to the Internet. IPv4 does have the capability for more than 4 billion addresses, but still is not adequate to handle the demand, not so much for the number of addresses it can handle, but because of the way it groups bits for its network/host numbering system. The problem here is that IPv4's numbering system wastes address assignments and suffers from excessive routing overhead.

A Basic Overview of IPv4

IPv4 supports a fixed 32-bit field for addressing, which is no longer sufficient for the number of users on the Internet. Routing tables are growing exponentially and this has been causing a great deal of difficulty for many organizations. In addition, autoconfiguration and scalable multicast are needed. Furthermore, there is a need to develop real-time flow for video conferencing, as we will be discussing throughout this book. These remain the key issues associated with the move toward a new protocol format.

IPv4 addresses are categorized according to the size of a network (numbers of IP addresses used). The categories are known as address classes. We are concerned with the first three categories; they maintain different number of bits for the networkID portions of their addresses. Class B, for example, has 14 bits for the networkID and 16 for the hostID combining to form 16,384 outcomes; each of these outcomes can accommodate 65,534 hosts.

The major problem with this scheme is that the most numerous networks are those in the middle of this class structure. The number of addresses has remained static and the distribution of networks has been evolving to lump around the intermediate networks. Subnettting and supernetting have been developed, in part, to help fix this problem.

The Addressing System of IPv4

IP addressing is based on network and host number assignment. In IPv4, these numbers are organized as 32-bit addresses, with host numbers and network numbers embedded in the addresses. These numbers identify the network or host connection and not the actual network or computer. IPv4 divides its address assignment into three main classes: A, B, and C.

- ✔ Class A addresses assign the first 7 bits (or 1 byte) to a network and the last 24 bits (or 3 bytes) to a host. These addresses are reserved for organizations with up to 16,777,216 hosts, and there can be at most 128 of these networks.
- ✔ Class B addresses assign the first 14 bits (or 2 bytes) to a network and the last 16 bits (or 2 bytes) to a host. These addresses are reserved for organizations with up to 65,536 hosts; there can be at most 16,384 networks.
- ✔ Class C addresses assign the first 21 bits (or 3 bytes) to a network and the last 7 bits (or 1 byte) to a host. These addresses are reserved for organizations with fewer than 256 hosts; there can be at most 2,097,152 networks.

The address class determines the network mask of the address. A network mask is a 32-bit Internet address that has all the bits in the network number set to one and all the bits in the host number set to zero. Hosts and routers use the network mask to route Internet packets.

The Address Management Issues

Although the amount of possible addresses seems enough to suit the world needs for IP addresses, the way IPv4 handles the addresses within each of these classes prevents this. For example, an organization seeks 300 host addresses. The number of IP addresses the organization seeks puts it into the Class B category. If the company is assigned a Class B address, it would have 65,536 hosts, significantly more than needed, which wastes about 65,000 addresses.

To avoid this type of situation, the Classless Inter-Domain Routing (CIDR) scheme was introduced. CIDR essentially eliminates the class structure of addressing and instead allows the assignment of network numbers at any bit boundary. In this way network numbers can be created, for example, by aggregating several contiguous class C addresses. CIDR requires that network masks be explicitly specified when needed, rather than allowing them to be implicitly derived from the address (as in the class system).

Another problem resulting from IPv4's address classes is the Internet backbone router table size explosion. CIDR also addresses this by allowing for address aggregation. However, a negative aspect to CIDR is that with an arbitrary address, the network and host numbers cannot be determined unless the network mask is known. The limitations of IPv4 have quickly been realized and measures such as CIDR have extended its life slightly. Worldwide network demand, however, is making the need for IPv6 immediate.

The Need for IPv6

It is anticipated that in the early 21st century the Internet will be routinely used in ways unfathomable to us today. Its usage is expected to extend to multimedia notebook computers, cellular modems, and even appliances such as TVs, toasters and coffee makers (remember that IBM's latest desktop PC model already comes with some of these remote functionalities to control your home appliances).

Virtually all the devices with which we interact, at home, at work, and at play, will be connected to the Internet—the possibilities are endless, and the implications staggering, especially as far as security and privacy go.

The advent of the IPv6 initiative does not mean that the technologies will exhaust the capabilities of IPv4, our Internet technology. However, there are still compelling reasons to begin adopting IPv6 as soon as possible. This process has its challenges, and it is essential to any evolution of Internet technology that there be requirements for seamless compatibility with IPv4, especially with regard to a manageable migration, which would permit taking advantage of the power of IPv6, without forcing the entire Internet to upgrade at once.

Thus IPv6 becomes a central point, the cornerstone for the Internet and its viability for corporate networks, IP multicasting, global e-commerce, and telephony applications such as voice over IP and more.

Some IPv6 Advantages

It is important to note the intent behind IPv6 design. It was not designed to be a huge leap away from what has worked in IPv4. This would be catastrophic, as backward compatibility would not be assured. This version of the protocol is designed to be an outward growth from something that works but no longer fits the requirements of the user community.

IPv6 Address Enhancements

IPv6 has kept all the working functionality of IPv4. Things that did not work well in IPv4 were intentionally left out of the new version. Below is a summary list of the main changes implemented in IPv6:

- ✔ Expanded Routing and Addressing Capabilities
- ✔ Increased IP address size from 32 bits to 128 bits, to support more levels of addressing hierarchy, a much greater number of addressable nodes, and simpler autoconfiguration of addresses

- ✔ Improved scalability of multicast routing, with a "scope" field added to multicast addresses
- ✔ A new type of address called a "anycast address", to identify sets of nodes where a packet sent to an anycast address is delivered to one of the nodes. The use of anycast addresses in the IPv6 source route allows nodes to control the path along which their traffic flows
- ✔ Simplified Header Format
- ✔ Dropped or optional IPv4 header fields to reduce the common-case processing cost of packet handling and to keep the bandwidth cost of the IPv6 header as low as possible, despite the increased size of the addresses. Even though the IPv6 addresses are four times longer than IPv4 addresses, the IPv6 header is only twice the size of the IPv4 header
- ✔ Improved Support for Options
- ✔ Changed encoding of IP header options to allow for more efficient forwarding, less stringent limits on the length of options, and greater flexibility for introducing new options in the future
- ✔ Quality-of-Service Capabilities
- ✔ Added capability to enable the labeling of packets belonging to particular traffic "flows" for which the sender requests special handling, such as non-default quality of service or "real- time" service
- ✔ Authentication and Privacy Capabilities
- ✔ Defined extensions, which provide support for authentication, data integrity, and confidentiality. This is included as a basic element of IPv6 and will be included in all implementations.

Autoconfiguration

In autoconfiguration (plug-and-play), introduced as a concept in IPv6, a host established as a resource on the Internet will be not be required to re-establish itself as a host. This host will be able to connect as a node on the network with a minimal amount of configuration. This will greatly reduce the time LAN administrators spend configuring and maintaining IP address leases. In a broader context, individuals who travel will not be required to reconfigure in order to gain connectivity to the Internet.

An added benefit of this aspect of IPv6 is that it does not require DHCP. A "local link IP address" will be developed upon the initialization of a physical layer device such as a NIC. As part of the Ethernet standard for example, such addresses are unique. Building on these addresses and creating a unique IP address as a derivation of the Ethernet address will ensure successful addressing of the NIC. This is the IP address that can be established in auto-configuration. This approach is similar to that of IPX, which has been quite successful.

IPv6 Header

The IPv6 header, as shown in Figure 1.1, consists of eight discrete items, many of them innovative and obviously directly targeted at some of the shortcomings of IPv4. These items are: Version, Prior(ity) Flow Label, Payload Length, Next Header, Hop Limit, Source Address and Destination Address.

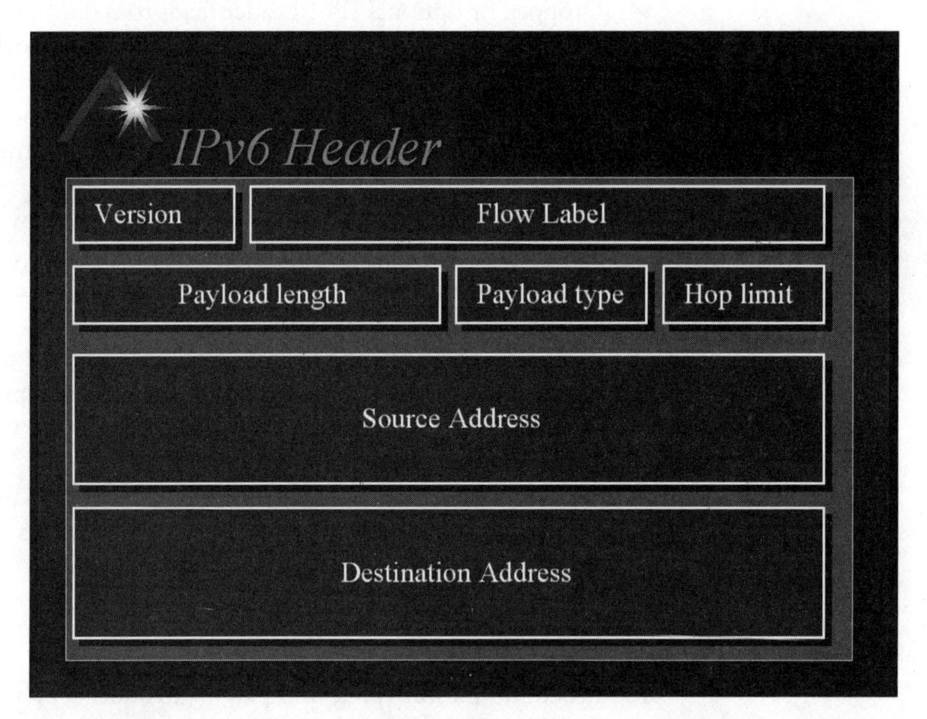

FIGURE 1.1 IPv6 header, broken down into eight discrete items

IPv6 Extensions

IPv6 includes an improved option mechanism. IPv6 options are placed in separate extension headers located between the IPv6 header and the transport-layer header in a packet. IPv6 extension headers are not examined or processed by any router along a packet's delivery path until it arrives at its final destination. This facilitates a major improvement in router performance for packets containing options. In IPv4 the presence of any options requires the router to examine all options.

The other improvement is that unlike IPv4 options, IPv6 extension headers can be of arbitrary length and the total amount of options carried in a packet is not limited to 40 bytes. This feature plus the manner in which they are processed, permits IPv6 options to be used for functions which were not practical in IPv4. A good example of this is the IPv6 Authentication and Security Encapsulation options.

In order to improve performance when handling subsequent option headers and the transport protocol which follows, IPv6 options are always an integer multiple of eight octets long, in order to retain this alignment for subsequent headers. The IPv6 extension headers currently defined are:

- ✔ **Routing**—Extended Routing (like IPv4 loose-source route)
- ✔ **Fragmentation**—Fragmentation and Reassembly
- ✔ **Authentication**—Integrity and Authentication
- ✔ **Security Encapsulation**—Confidentiality
- ✔ **Hop-by-Hop Option**—Special options which require hop-by-hop processing
- ✔ **Destination Options**—Optional information to be examined by the destination node.

Security Enhancements

IPv4 has a number of security problems and lacks effective privacy and authentication mechanisms below the application layer. IPv6 remedies these shortcomings by having two integrated options that provide security services. These two options may be used singly or together to provide differing levels of security to different users. This is important because different user communities have different security needs.

The first mechanism, called the *IPv6 Authentication Header*, is an extension header which provides authentication and integrity (without confidentiality) to IPv6 datagrams. While the extension is algorithm-independent and will support many different authentication techniques, the use of keyed MD5 is specified as the default algorithm to help ensure interoperability within the worldwide Internet. This can be used to eliminate a significant class of network attacks, including host masquerading attacks.

The use of the IPv6 Authentication Header is particularly important when source routing is used with IPv6 because of the known risks in IP source routing. Its placement at the Internet layer can help provide host-origin authentication to those upper layer protocols and services that currently lack meaningful protections. This mechanism should be exportable by vendors in the United States and countries with similar export restrictions because it only provides authentication and integrity, and specifically does not provide confidentiality. The exportability of the IPv6 Authentication Header encourages its widespread deployment and use.

The second security extension header provided with IPv6 is the *IPv6 Encapsulating Security Header*. This mechanism provides integrity and confidentiality to IPv6 datagrams. It is simpler than some similar security protocols (e.g., SP3D, ISO NLSP) but remains flexible and algorithm-independent. To achieve interoperability within the global Internet, DES CBC is being used as the standard default algorithm with the IPv6 Encapsulating Security Header.

Transitioning to IPv6

The key transition objective is to allow IPv6 and IPv4 hosts to interoperate. A second objective is to allow IPv6 hosts and routers to be deployed in the Internet in a highly diffuse and incremental fashion, with few interdependencies. A third objective is that the transition should be as easy as possible for end-users, system administrators and network operators to understand and carry out.

The IPv6 transition mechanisms are a set of protocol mechanisms implemented in hosts and routers, along with operational guidelines for addressing and deployment, designed to make transition the Internet to IPv6 work with as little disruption as possible. These transition mechanisms provides a number of features, including:

- ✔ **Incremental upgrade and deployment**—Individual IPv4 hosts and routers may be upgraded to IPv6 one at a time without requiring any other hosts or routers to be upgraded at the same time. New IPv6 hosts and routers can be installed one by one.
- ✔ **Minimal upgrade dependencies**—The only prerequisite to upgrading hosts to IPv6 is that the DNS server must first be upgraded to handle IPv6 address records. There are no prerequisites to upgrading routers.
- ✔ **Easy Addressing**—When existing installed IPv4 hosts or routers are upgraded to IPv6, they may continue to use their existing addresses. They do not need to be assigned new addresses. Administrators do not need to draft new addressing plans.
- ✔ **Low start-up costs**—Little or no preparation work is needed in order to upgrade existing IPv4 systems to IPv6, or to deploy new IPv6 systems. The mechanisms employed by the IPv6 transition mechanisms include:
  - An IPv6 addressing structure that *embeds IPv4 addresses within IPv6 addresses*, and encodes other information used by the transition mechanisms

- A model of deployment where all hosts and routers upgraded to IPv6 in the early transition phase are "dual" capable (i.e. implement complete IPv4 and IPv6 protocol stacks)
- A technique of encapsulating IPv6 packets within IPv4 headers to carry them over segments of the end-to-end path where the routers have not yet been upgraded to IPv6
- A header translation technique to allow the eventual introduction of routing topologies that route only IPv6 traffic, and the deployment of hosts that support only IPv6. Use of this technique is optional, and occurs in the later phase of transition if it is used at all.

The IPv6 transition mechanisms ensure that IPv6 hosts can interoperate with IPv4 hosts anywhere in the Internet until the time when IPv4 addresses run out, and allows IPv6 and IPv4 hosts (within a limited scope) to interoperate indefinitely after that. This feature protects the huge investment users have made in IPv4 and *ensures that IPv6 does not render IPv4 obsolete*. Hosts that need only a limited connectivity range (e.g., printers) need never be upgraded to IPv6.

The 6bone Initiative

This project is actually much more important than one would think from its name. It is essentially a practice ground for learning more about the use of IPv6 as well as for fostering the implementation of the new standard.

The project is a close relative of the Internet Engineering Task Force (IETF) and currently spans three continents. One of its main purposes is to develop and implement a backbone (thus the name, we suppose) able to support IPv6. A new protocol is not worth much if support is unavailable. The thinking is that eventually the backbone will mimic the structure that exists today in that it will consist of ISPs as well as other networks combined to provide a great deal more functionality and power to the Internet.

The problem that 6bone answers is an important one: how can we test new functionality without placing the existing functionality at risk? This project involves placing a virtual network layer on *top* of physical IPv4 network layers. The particulars of this setup can be found at http://www.6bone.com as they are beyond the scope of this book. It is important to note however, that as IPv6 is adopted and implemented, 6bone will eventually be phased out.

6bone is interested in developing policy and procedures for the next stage of IP integration. It is not designed to develop new network architectures or infringe upon the way in which networking is accomplished by major players in the Internet. The project attempts to include as many large players as possible in order to develop policies and procedures that can be adopted by many organizations.

As we have learned, there are a number of changes in store for the protocol suite that are important to understand and be prepared for. These changes are not designed to create a huge lead away from what has worked to date, but *are* designed to build upon the aspects of IP that have worked and move away from those that have not.

IPv4 is a solid, routable protocol. In order for larger network environments to use this product, they require some sort of connectivity that is usually filled by a DHCP (Dynamic Host Configuration Protocol) and DNS (Dynamic Name Server) server. IPv6 has the potential to circumvent many of these requirements and provides the opportunity to create a more efficient, secure networking environment.

IPv6 has been designed to enable high-performance, scalable internetworks to remain viable well into the next century, and by so doing, corrected many inadequacies of IPv4 (see Figure 1.2 for a sample IPv6 packet). But in order to fully take advantage of IPv6 improvements its full spectrum of benefits must be used. Some of these are found in obviously enhanced features, others are less tangible and relate to the fresh start that IPv6 provides to LAN and Internet administrators.

FIGURE 1.2
Sample IPv6 packet

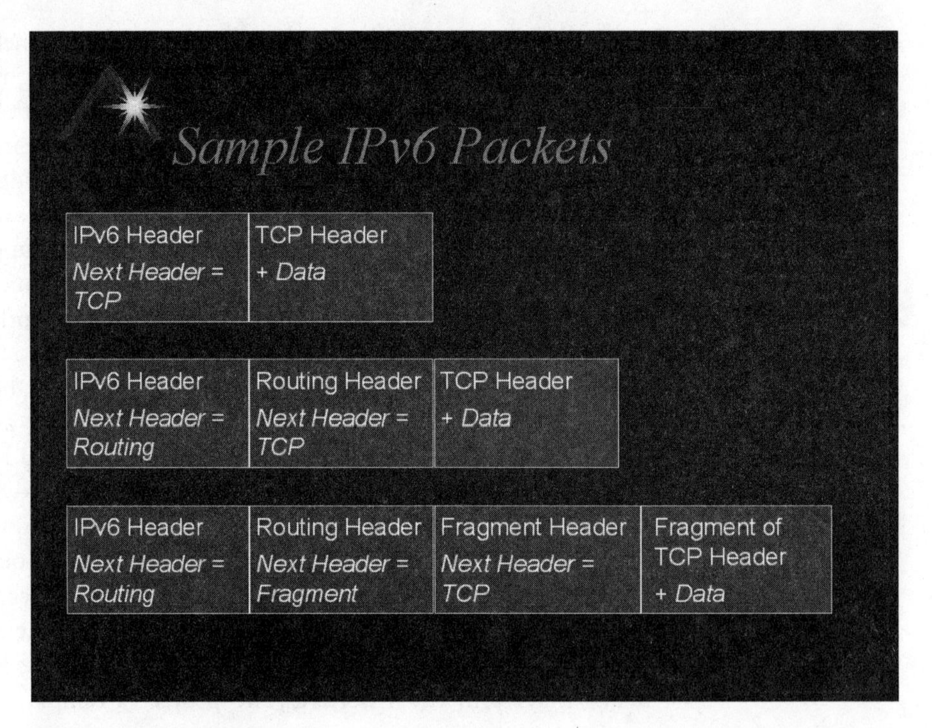

Addressing and Routing

IPv6 provides a framework for solving some critical problems that currently exist inside and between enterprises, as shown in Figure 1.3. IPv6 will allow Internet backbone designers to create a highly flexible and open-ended global routing hierarchy. At the level of the Internet backbone where major enterprises and Internet Service Provider (ISP) networks come together, it is necessary to maintain a hierarchical addressing system, much like that of the national and international telephone systems. Large central-office phone switches, for instance, only need a three-digit national area code prefix to route a long-distance telephone call to the correct local exchange. Likewise, the current IPv4 system uses a (somewhat haphazard) form of address hierarchy to move traffic between networks attached to the Internet backbone.

FIGURE 1.3
IPv6 addressing enhancements

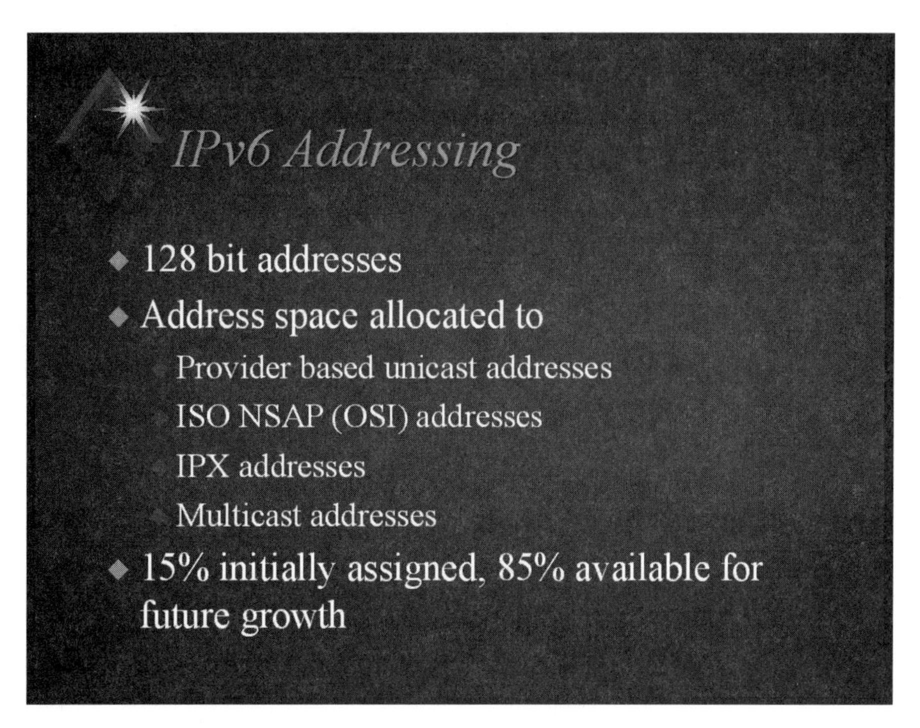

Without an address hierarchy, backbone routers would be forced to store routing table information on the reachability of every network in the world. Given the current number of IP subnets in the world and the growth of the Internet, this is not feasible. With a hierarchy, backbone routers can use IP address prefixes to determine how traffic should be routed through the backbone. IPv4 uses a technique called CIDR, which allows flexible use of variable-length network prefixes. With this flexible use of prefixes, CIDR permits considerable "route aggregation" at various levels of the Internet hierarchy, which means backbone routers can store a single routing table entry that provides reachability to many lower-level networks.

The availability of CIDR routing does not guarantee an efficient and scalable hierarchy. In many cases, legacy IPv4 address assignments that originated before CIDR do not facilitate summarization. These issues affect high-level service providers and individual end users in all types of businesses. Figures 1.4 and 1.5 outline some of the main features of IPv6 routing.

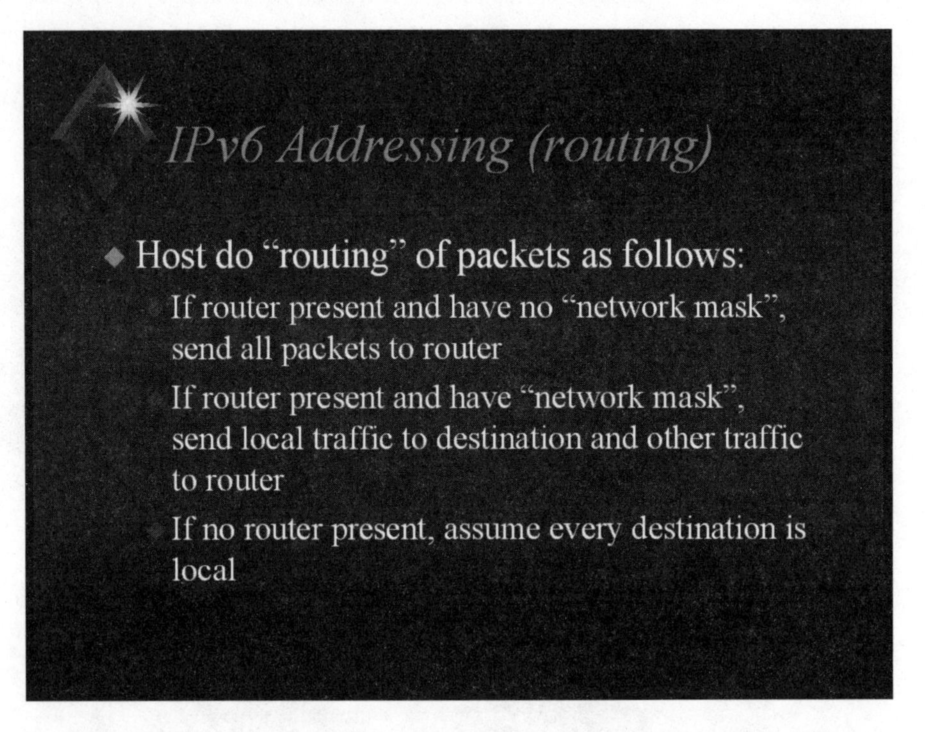

FIGURE 1.4 IPv6 addressing routing features

Gateways and network address translators typically limit outside-world connectivity for users in private address spaces with non-unique addresses. NAT (Network Address Translators) services are meant to allow an enterprise to have whatever internal address structure it desires, without concern for integrating internal addresses with the global Internet. The NAT device sits on the border between the enterprise and the Internet, converting private internal addresses to a smaller pool of globally unique addresses that are passed to the backbone and vice versa.

NAT may be appropriate in some organizations, particularly if full connectivity with the outside world is not desired. Figure 1.6 gives an example of the NAT layout. But for enterprises that require robust interaction with the Internet, NAT devices are not always desirable. The NAT technique of substituting address fields in each packet that leaves and enters the enterprise is very demanding, and can lead to a bottleneck between the enterprise and the Internet.

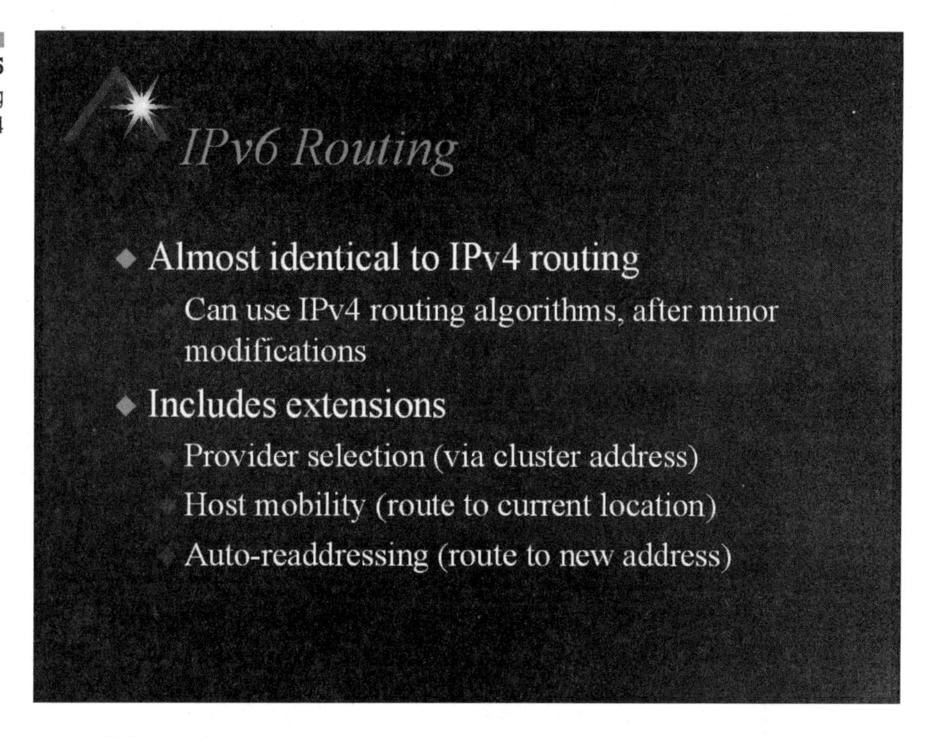

FIGURE 1.5 IPv6 routing resembles IPv4

An NAT may keep up with address conversion in a small network, but as Internet access increases, the NAT's performance must increase in a parallel fashion. The bottleneck effect is exacerbated by the difficulty of integrating and synchronizing multiple NAT devices within a single enterprise. It is highly unlikely that an enterprise will achieve the reliable high-performance Internet connectivity with NAT that is common today with multiple routers attached to an ISP backbone in an arbitrary mesh fashion.

Another limitation of IPv4 relates to the ongoing need in many organizations to renumber stations. When an enterprise changes ISPs, it may have either to renumber all addresses to match the new ISP-assigned prefix, or to implement address translation devices. Renumbering is also a reality for many corporations that undergo a merger or an acquisition that entails network consolidation. Also, address shortages and routing hierarchy problems are increasingly a threat to the network operations of larger (and to some extent small) enterprises. Smaller networks can be completely dropped from Internet backbone routing tables if they do not adhere to the address hierarchy. In the current system, ISPs with individual dial-in clients cannot allocate IP numbers as freely as they wish. Consequently, many dial-in users must use an address allocated from a pool on a temporary basis. In other cases, small dial-in sites are forced to share a single IP address among multiple end systems.

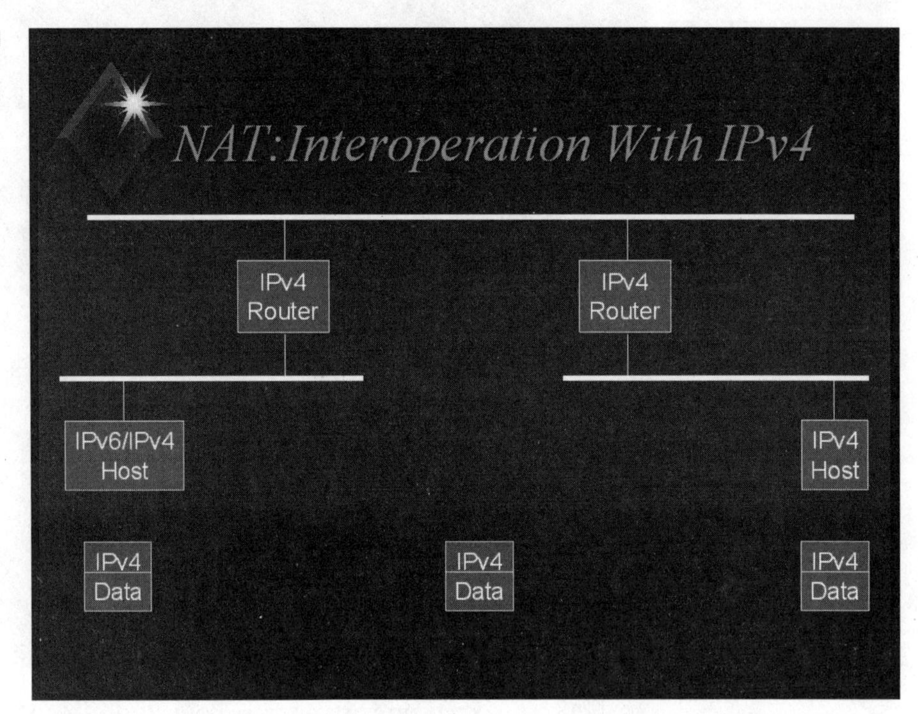

FIGURE 1.6 NAT allows interoperation with IPv4

IPv6 for Businesses

Today there are plenty of business issues encouraging IPv6, from protocol basics to industry realities and demands. For one thing, if we look at the increasing business requirements for interactive multimedia (IP multicasting and voice over IP included), and high-bandwidth network applications, IPv6 is critical to the continued viability of enterprise internetworks and the public Internet. To say that IPv6 is driven by a need to expand Internet addresses is not only simplistic, but misses all the potential behind the whole IPv6 initiative. Figure 1.7 lists some, but not all, of the main reasons driving the development of IPv6.

For many, IPv6 is a proposed solution for preventing IPv4's obsolete 32-bit address space from running out of network layer addresses, as we discussed earlier in this chapter. But the Internet Network Information

Center (InterNIC), the authority that assigns blocks of IP addresses to large network service providers and network operators, predicts that IPv4's address format should exhaust about halfway into the next decade.

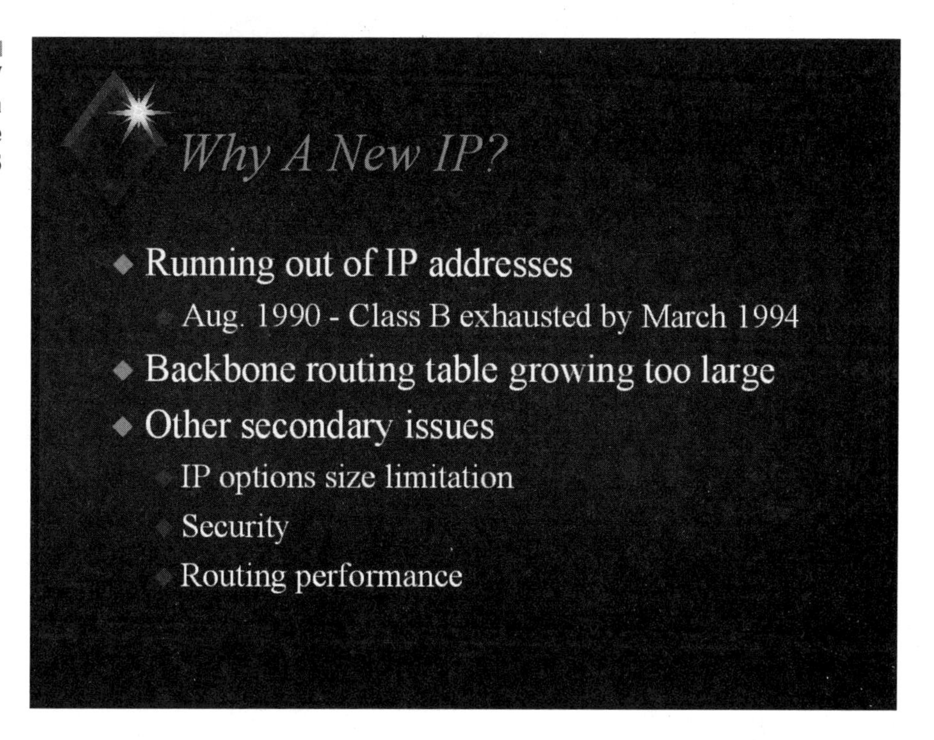

FIGURE 1.7 Some of the main reasons driving the development of IPv6

The fact is that IPv6 has great advantages over Ipv4, and consequent impact. Consider this: IPv6's space address jumps from Ipv4's 32-bit format to a 128-bit address space. Figures 1.8 and 1.9 outline some of the main changes in IPv6 from IPv4. This new capability alone should be more than enough to grant unique addresses for every conceivable variety of network device in the world for many decades to come. More precisely, its 16-byte addressing capability can handle up to 340,282,366,920,938,463,463,374,607,431,768,211,456 IP addresses. Although a major achievement, the features of IPv6 address several issues, for good and for bad. It is important to be aware of and understand both advantages and disadvantages, better to understand the impact of IPv6 on business and the internetworking environment.

FIGURE 1.8 Some of the main changes of IPv6 from IPv4

FIGURE 1.9 More changes of IPv6 from IPv4

The expanded IP addressing of IPv6 gets a lot of attention but is only one of many important features. Figure 1.10 lists other new features of IPv6, but as mentioned earlier, IPv6 also addresses important critical business requirements, as follows:

- ✔ Increased scalability for network architectures
- ✔ Improved security
- ✔ Improved data integrity
- ✔ Integrated quality-of-service (QoS)
- ✔ Autoconfiguration
- ✔ Mobile computing features
- ✔ Data multicasting
- ✔ More efficient network route aggregation at the global backbone level.

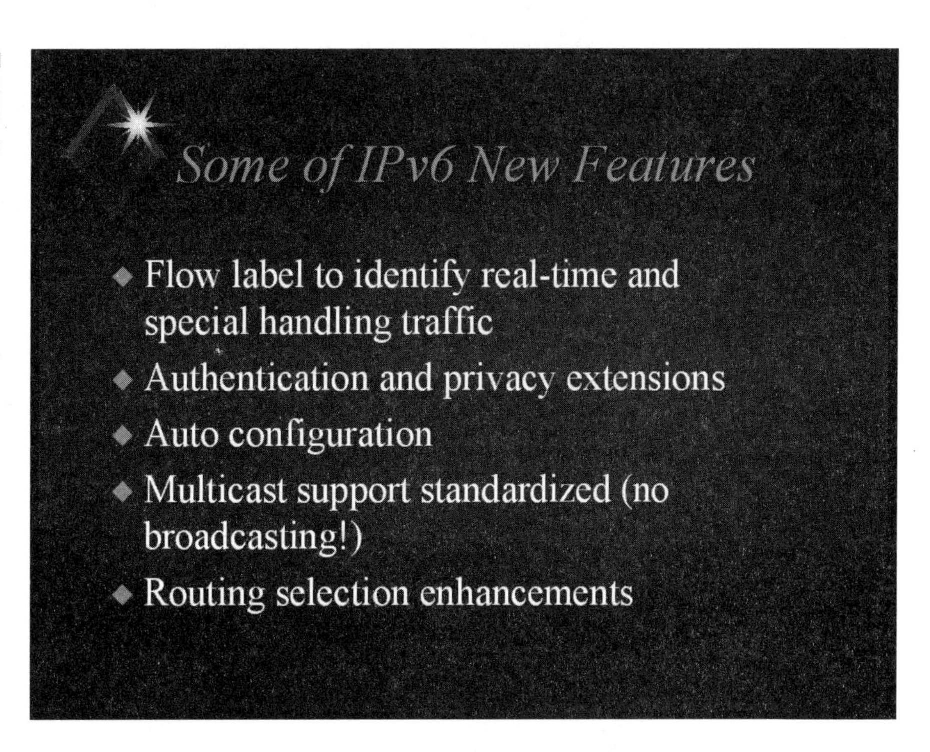

FIGURE 1.10 Some new IPv6 features

Do not be deceived by the list of features above. Just as everything has a price in this world, these benefits of IPv6 also have a price, which can be viewed as a disadvantage, even though a temporary one, as the advantages will not come without a transition effort. IPv6 presents such challenges that many in the industry defend the idea of extending the life of IPv4 indefinitely with changes to the protocol standards and various proprietary techniques.

Figure 1.11 outlines major challenges facing those migrating onto IPv6: the lack of a finished product and not being PnP (Plug-and-Play). Another challenge occurs in the case of the NATs, which preserve IPv4 address space by intercepting traffic and converting private intra-enterprise addresses into globally unique Internet addresses. The many quality-of-service and security enhancements to IPv4 also are being extended.

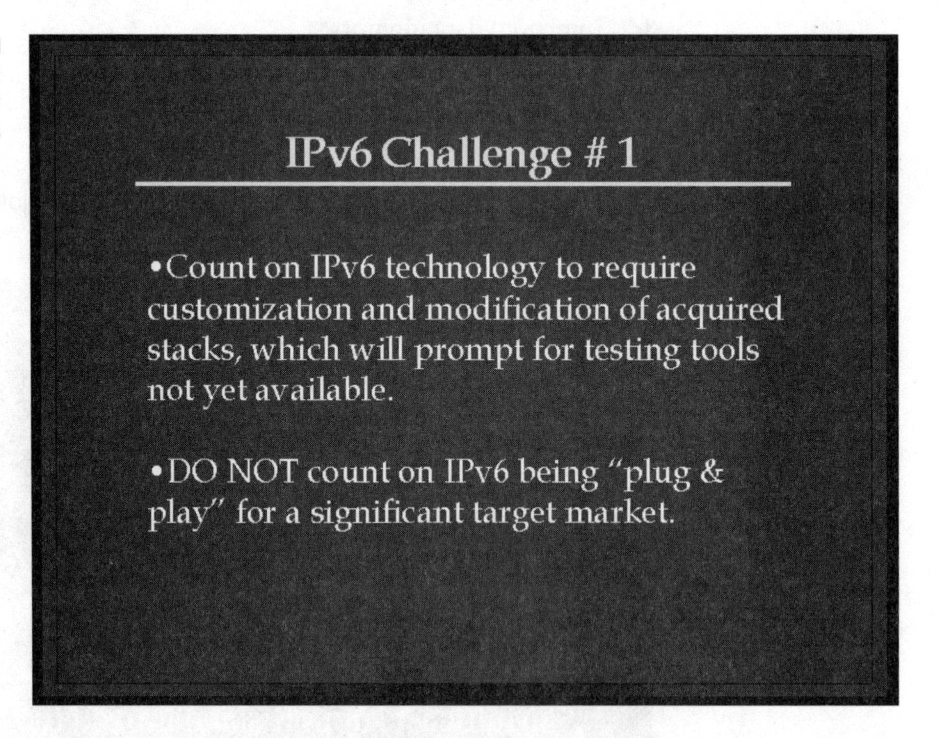

FIGURE 1.11 Major challenges for early IPv6 migrators

Some professionals believe that the wide applicability of IPv6 support is not something to involve and concern business and end-users in general. But we can be sure that IPv6 will connect more than computers and networks. IPv6 will make it possible to connect many other devices, such as palmtop personal data assistants (PDA), hybrid mobile phone embedded network components, as well as other devices and equipment that will range from coffee makers to sprinklers at a golf course.

These factors (shown in Figure 1.12), exemplify some of the main challenges faced by IPv6 during its earlier implementation phase. Furthermore, as new devices make their way onto the Internet, they will strain the existing network fabric in ways the early IP protocol designers could hardly have imagined. IPv6's 128-bit address space will allow new markets to deploy an enormous array of new applications and devices for desktop, mobile, and embedded networks with very high return of investment (ROI). Moreover, be confident that these new trends will be led and pushed by end-user applications. The heavily competitively networked business environment of the new century will have to exploit all the capabilities of IPv6, not only to create a highly scalable address space, but also to take advantage of its strong capability for autoconfiguration services, capital for large networks.

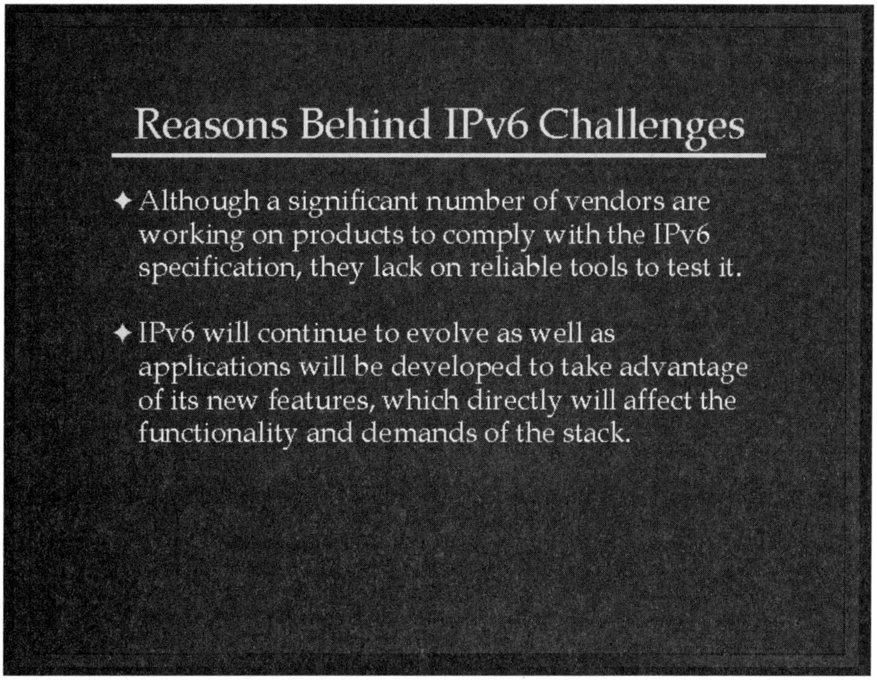

Reasons Behind IPv6 Challenges

- Although a significant number of vendors are working on products to comply with the IPv6 specification, they lack on reliable tools to test it.
- IPv6 will continue to evolve as well as applications will be developed to take advantage of its new features, which directly will affect the functionality and demands of the stack.

FIGURE 1.12 Causes responsible for the challenges presented to IPv6 during its early implementation

Relevance of IPv6

Very soon we will be able to use our car's on-board computer to take dictation while we are stuck in traffic, update our schedule so that family and colleagues know where we are, and instruct our house to turn up the heat and switch on the lights. That same car will automatically run diagnostics and download the results, so when we get to the shop the needed parts are ready and waiting. Virtually all the devices with which we interact, at home, at work, and at play, will be connected to the Internet—the possibilities are endless, and the implications staggering. IPv6 will make all this possible.

IPv6 is already promoting major advantages to backbone routers, enabling efficient multitiered routing hierarchies that limit the uncontrolled growth of backbone router tables. It is also benefiting end-users, as it enables them to run more secure Intranet environments since IPv6 offers encryption and authentication services as an integral part of its IP stack. The advantages are even greater for mobile users, always moving from one location to another with their notebook computers (and palmtops). It also brings major advantages for other dynamic departmental staffs, such as teams of auditors and inspectors conducting due diligence examinations outside their offices; IPv6 enables the use of automatic configuration in the assignment of IP addresses. The frustration of manual administration of IP addresses is over with IPv6, as are the time consumption and costs associated with that administration.

Although the explosive growth of the World Wide Web and other, more futuristic, technologies will not totally exhaust the capabilities of our current generation of Internet technology until early in the next decade, there are still compelling reasons to begin adopting IPv6 now. Essential to any evolution of Internet technology are the requirements for seamless compatibility with current technology (IPv4), and manageable migration. Thus, it will be possible to take advantage of the power of IPv6 now, where needed, without forcing the entire Internet to upgrade simultaneously.

We should clarify a misconception when considering IPv6 with reference to Asynchronous Transfer Mode (ATM) cell switching, and other switching methods, as possible replacements for packet routing. ATM has its place in the internetworking industry, but cannot replace packet routing by itself. Thus, it is not a matter of choice to have ATM or IPv6 because the two protocols not only complement each other, but also serve entirely different roles in corporate networking. As a matter of fact, why not use ATM as a transmission medium for high-speed IPv6 backbone networks? This is a question that has triggered standard and developmental work aimed at the integration of ATM and IPv6. As discussed in more detail in the chapters to come, especially Chapter 3, IPv6, just like IPv4, provides network-layer services over all major link types, including ATM, Ethernet, Token Ring, ISDN, Frame Relay, and T1.

IPv6 Multicasting

The designs of current network technologies were based on the premise of one-to-one, or one-to-all communications. This means that applications distributing information to a large number of users must build separate network connections from the server to each client. IPv6 provides the opportunity to build applications that make much better use of server and network resources through its multicasting option. This allows an application to broadcast data over the network; only those clients receive it who are properly authorized to do so. Multicast technology opens up a whole new range of potential applications, from efficient news and financial data distribution, to video and audio distribution, etc.

The Conversion Challenge

While a primary design goal of IPv6 is to ease the transition from and coexistence with IPv4, converting today's tens of millions of IPv4-based systems to IPv6 will be a major challenge. However, IPv6's built-in compatibility features will ease the pain, and options like tunneling IPv4 packets over IPv6, tunneling IPv6 packets over IPv4, and translation gateways will help make the job easier.

Many organizations are working on IPv6 drivers for the popular UNIX BSD operating environment. Network software vendors have announced a wide range of support for IPv6 in network applications and communication software products.

Changes to protocol systems can have profound effects on existing applications and must be carefully implemented to minimize risk. Thus, migrating from IPv4 to IPv6 in existing applications, or implementing IPv6 in new applications, requires considerable expertise to ensure a smooth transition and trouble-free implementation.

Business Opportunities for IPv6

Vendors operating in the IP industry should be attentive to new opportunities in the IPv6 market, as I already anticipate market dynamics, which will lead to universal IPv6 adoption, as discussed earlier in this chapter. In order to be successful, vendors should look for product concepts which leverage these market dynamics.

One market to tap is the fact that the IPv6 stack is incompatible with IPv4 stack. The IETF is formalizing two approaches to the migration process: tunneling and dual stacks, which are discussed later in this book.

Tunneling, as shown in Figure 1.13, is an alternative, as it addresses the incompatibility of the infrastructure—by enabling two IPv6 nodes to communicate over an IPv4 backbone (or vice versa), but it does not enable an IPv4 node to communicate with an IPv6 node. What if a node with a dual

IPv6/v4 stack, which could communicate with either IPv6 or IPv4 nodes, were available? Would dual stacks resolve the migration issues?

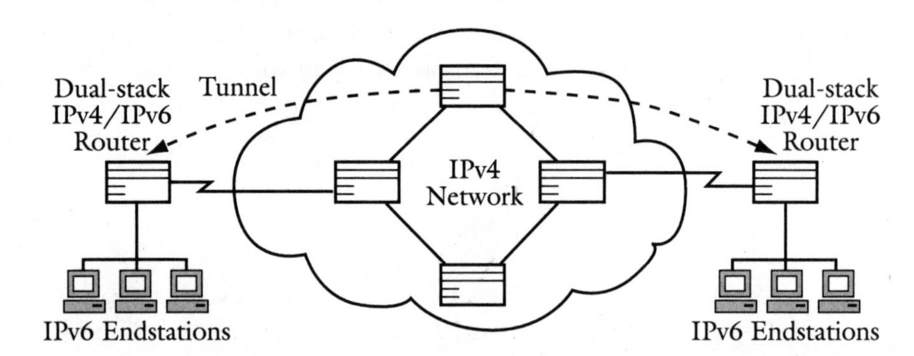

FIGURE 1.13 The transmission of IPv6 packets can be made over IPv4 via tunneling

If so, would it be advantageous to install a dual stack instead of IPv4 only? Clearly this would require a considerable investment, especially at a large site, and coordination of stacks from different vendors would be necessary. Also, upper-level protocols would need to be replaced, and a "separate but equal" addressing and configuration scheme would need to be implemented and supported. The question is, why incur these costs and risks? The possible answers could be:

1. It would be desirable to have some target nodes only accessible via IPv6. But why do that when such a choice severely constrains communicating nodes within an organization?
2. The compelling features of IPv6 are many. It may be worthwhile to endure the challenges brought by IPv6, as Microsoft users endure the beta versions of Windows 98 and NT 5.0, for the attributes of IPv6, such as quality of service, automatic configuration, security and large address space. But these latter attributes will also available via IPv4 or have adequate IPv4 workarounds for the next decade.

This is an IPv6 constraint. The advantages of dual stack installation may not warrant the costs and risks. They do however, pose a "chicken or egg" conundrum for early IPv6 adopters.

There is an alternate approach that may mitigate this problem. What about a device (or scheme) that translates between IPv4 and IPv6 protocols, as shown in Figure 1.14? IPv4 nodes, which need to communicate with IPv6 nodes, would not require any stack upgrades. New sites might consider IPv6 installations without fear of incompatibility with the installed base. IPv6 migration could proceed based on its larger address space.

FIGURE 1.14 Using a translation scheme to bridge IPv6 and IPv4 packets

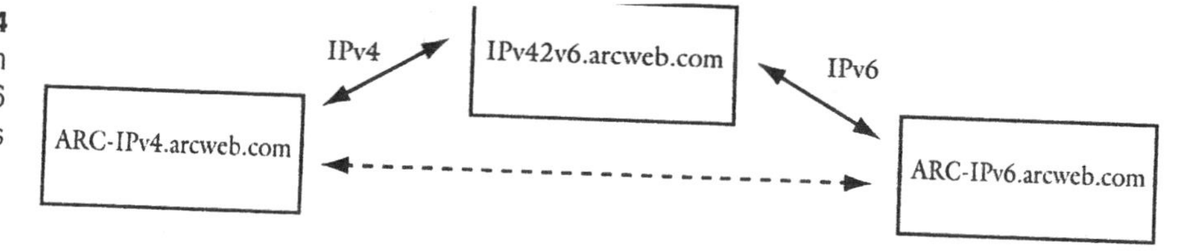

IPv6's Unicast, Multicast and Anycast Addressing

IPv6's unicast addresses identify a single interface. Packets sent to a unicast address are delivered to the interface identified by that address. There are four types of unicast addresses:

- ✔ Provider-based, allowing global addressing to all connected hosts
- ✔ Local-use–link-local for addressing on a single link (physical network) or subnetwork, and site-local, designed for local use that can later be integrated into global addressing
- ✔ IPv4-compatible, which provides compatibility between IPv4 and IPv6 until a complete transition is attained
- ✔ Loopback, which sends an IPv6 packet to itself. These packets are not sent outside a single node.

Multicast addresses identify a set of interfaces that usually belong to different nodes. Packets sent to a multicast address are delivered to all interfaces identified by that address. This is useful in several ways, such as sending discovery messages only to the machines that are registered to receive them. A particular multicast address can be confined to a single system, restricted to a specific site, associated with a particular network link, or distributed worldwide.

Note that IPv6 has no broadcast addresses and uses multi[illegible] instead, as sh[illegible]

TABLE 1.1 A multicast address allows multiple configurations: being part of a single system, restricted to a specific site, associated to a network link or distributed worldwide

Bits	8	4	4	112
	11111111	Flags	Scope	Group ID

Anycast addresses are a new introduction to IP technology with the IPv6 protocol. This kind of address identifies a set of interfaces, usually belonging to different nodes. A packet sent to an anycast address is delivered to one of the interfaces identified by the address. This is usually the nearest interface, and is determined by how the router measures distance.

This makes routing more efficient, as shown in Table 1.2, because the address itself can specify intermediate hops en route to a destination, rather than having the router determine the route.

TABLE 1.2 Anycast addresses enable more efficient routing by specifying intermediate hops en route to a destination

Bits	N	128-n
	Subnet prefix	0000 000

Address Resolution and Neighbor Discovery

In order for Internet packets to be transferred in a particular subnet on a particular medium, the nodes need to know the subnet address or the media address of the target station. IPv4 relies on the Address Resolution Protocol (ARP), but IPv6 uses *neighbor discovery*, which provides the same resources as ARP but also adds router discovery.

[illegible] of IPv6 ICMP how IPv6 uses multicast transmission to identify the media address of a destination. The message will always be sent to multicast addresses every time the media address of the destination is unknown. Thus, for IEEE-802, Ethernet or FDDI, the 48-bit multicast address is obtained by concatenating a fixed 16-bit prefix, 3333, and the last 32 bits of the IPv6 multicast address. Xerox reserved this prefix for use with IPv6, as shown in Figure 1.15.

FIGURE 1.15 Mapping of a multicast address in IEEE-802 networks

33	33	DST13	DST14	DST15	DST16

RFC 1970, "Neighbor Discovery for IP Version 6 (IPv6)," specifies the standards-track protocol for the Internet community. As mentioned earlier, nodes and hosts need a mechanism to determine the address of a target node or host. Neighbor discovery enables this by a link-layer address process, for neighbors known to reside on attached links. Neighbor discovery is also used by hosts to find neighboring routers willing to forward packets. It also actively keeps track of which neighbors are reachable and which are not, and detects changed link-layer addresses. When a router or the path to a router fails, a host actively searches for functioning alternates.

TIP

For more details about the parameters of this specification check **http://playground.sun.com/pub/ipng/html/ipng-main.html**.

Besides solving many problems related to the interaction of nodes attached to the same link IPv6 also defines mechanisms for resolving other specific problems:

- ✔ **Router Discovery** resolves how hosts locate routers residing on attached links
- ✔ **Prefix Discovery** resolves how hosts discover the set of address prefixes that define which destinations are on-link for an attached link
- ✔ **Parameter Discovery** solves how a node learns the necessary link parameters, such as the link MTU, or Internet parameters, such as the hop-limit value to place on outgoing packets
- ✔ **Address Autoconfiguration** solves how nodes automatically configure an address for an interface
- ✔ **Address Resolution** resolves how nodes determine the link-layer address of an on-link destination, such as a neighbor, given only the destination's IP address
- ✔ **Next-hop Determination** is the algorithm used for mapping an IP destination address into the IP address of the neighbor to which traffic for the destination should be sent. The next-hop can be a router or the destination itself
- ✔ **Neighbor Unreachability Detection** defines how nodes will determine if a neighbor is no longer reachable. If neighbor nodes are used as routers, alternate default routers can be tried, but in the case of both routers and hosts, address resolution can be performed again
- ✔ **Duplicate Address Detection** resolves how a node determines if the address it wants to use is already in use by another node
- ✔ **Redirect** defines how a router informs a host of a better first-hop node to reach a particular destination

- ✔ **Neighbor Discovery** defines five different ICMP packet types:
  - A pair of Router Solicitation and Router Advertisement messages
  - A pair of Neighbor Solicitation and Neighbor Advertisement messages
  - A Redirect message.

The messages serve the following purpose:

- ✔ **Router Solicitation**—When an interface becomes enabled, hosts may send out router solicitations that request routers to generate router advertisements immediately rather than at their next scheduled time
- ✔ **Router Advertisement**—Routers advertise their presence together with various link and Internet parameters either periodically, or in response to a router solicitation message. Router advertisements contain prefixes used for on-link determination and/or address configuration, a suggested hop-limit value, etc.
- ✔ **Neighbor Solicitation**—This is sent by a node to determine the link-layer address of a neighbor, or to verify that a neighbor is still reachable via a cached link-layer address. Neighbor solicitations are also used for duplicate address detection
- ✔ **Neighbor Advertisement**—This is a response to a neighbor solicitation message. A node may also send unsolicited neighbor advertisements to announce a link-layer address change
- ✔ **Redirect**—This is used by routers to inform hosts of a better first hop for a destination.

On multicast-capable links, each router is capable of periodically multicasting a router advertisement packet to announce its availability. In order to keep track of available routers, a host receives router advertisements from all routers, building a list of default routers. These router advertisements are frequently generated, so hosts can be updated about their presence every few minutes. However, hosts do not receive enough announcements from the routers to be able to rely on an absence of advertisements to detect router failure. For that, a separate neighbor unreachability detection algorithm provides failure detection.

IPv6's Multimedia Features

IPv6 incorporates a variety of functions that make it possible to use the Internet to deliver video and other real-time data that require guaranteed bandwidth and latency to ensure that packets arrive on a regular basis.

IPv6's Multicasting

IPv6 mandates support for multicast, a function that delivers messages to all hosts that register to receive it. This function makes it possible to deliver data simultaneously to large numbers of users for public or private consumption without wasting bandwidth broadcasting to the entire network. IPv6 also includes facilities to limit the scope of multicast message distribution to a specific location, region, company, etc., thereby reducing bandwidth usage and providing security.

Bandwidth Reservations

Using the mandated RSVP functionality, users can reserve bandwidth along the route from source to destination. This makes it possible to provide video or other real-time data with a guaranteed quality of service.

Packet Prioritizing

Packets will be assigned a priority level, ensuring that lower priority packets do not interrupt real-time data flow.

Jumbograms

IPv6 will support packet sizes of up to 4 billion bytes. This will make the transmission of large packets easier and ensure that IPv6 will be able to make the best use of all available bandwidth over any transmission media.

IPv6's Plug-and-Play Features

Currently, users or network managers must manually configure each machine with its address and other network information. This is a confusing, error-prone task for many individual users and a time-consuming chore for network managers. It also requires that to change network addresses, every machine must be manually reconfigured. IPv6 solves these problems by including mechanisms to allow hosts to discover their own addresses and to automate address changes.

Address Discovery

IPv6 allows hosts to learn their own addresses from a local router during boot-up, eliminating the need manually to configure addresses on each host. IPv6 also specifies procedures for a host to allocate an address for local site communications and for small sites without routers.

Network Information Discovery

IPv6 mandates support for DHCP which allows the host to obtain all relevant network information from a local router during boot-up.

Automated Address Changes

Because the router in IPv6 distributes network addresses, changing the address of the network requires only updating the router. In addition, all addresses include lifetimes, enabling the router to specify a time to switch addresses, insuring a smooth, error-free transition to a new address.

Support for Mobile Hosts

IPv6 will incorporate algorithms for automatic forwarding of packets from a base address to any other address. This will allow users connected to the Internet from any location, even mobile phones, and seamlessly receive their messages.

Dead Neighbor Detection

IPv6 specifies dead neighbor and dead gateway detection algorithms insuring that all implementations of IPv6 can efficiently detect problems and reroute packets when necessary.

IPv6 also enables applications to specify how to treat unknown options. This provides IPv6 with the flexibility to add new options in the future without necessitating that those existing implementations all be updated to conform.

Control Over Routing

As opposed to the capability to choose only loose (automatically determined) or strict (user specified) routing for the entire path in IPv4, in IPv6 users can specify loose or strict routing for each hop along the path. IPv6 also includes the flexibility to include additional routing methods in the future.

Configurability of Features

The IPv6 protocol for hosts and routers to discover neighboring machines is called neighbor discovery. IPv6 allows all the features of neighbor discovery such as retries and time-out parameters to be locally configured. This provides increased flexibility as well as the capability to optimize neighbor discovery for the needs and constraints of individual networks.

IPv6 Performance Considerations

Network performance is directly related to routing. The amount of traffic that leaves the local network (external traffic) compared to the amount of traffic that occurs on the network is constantly increasing. This is due in part to the demand for more services, especially graphics-based services. Speeds for LANs and WANs have also increased to hundreds of megabits per second, with gigabit networks not far in the future. Routers need to perform their functions of processing and forwarding IP datagrams much more quickly than before.

There are fewer fields in an IPv6 packet header than existed in IPv4. To increase the speed at which a packet travels past a router, separate optional headers are placed between the IPv6 header and the transport-layer header. Most of these are not examined or processed by routers along the packet's path, which simplifies and speeds up router processing. Additional optional headers are also easier to add, making IPv6 more flexible than IPv4. Because the IPv6 packet header has a fixed length, processing is also simplified.

IPv6 does not fragment packets as they are routed, as IPv4 does. Instead, packet fragmentation and reassembly will be done exclusively in the communicating hosts, thus reducing the workload for intermediate routers. When the transition to IPv6 is complete, the Internet will consist only of networks with Maximum Transmission Units (MTUs) equal to or larger than 576 bytes.

Performance with IPv6 will be optimized by the use of flow labels. The flow source specifies in the label any special service requirements from routers along a path, such as priority, delay, or bandwidth. All packets in the sequence carry the same details of this information in the flow label, to reserve the type of service they need from intermediate routers. Such a need might be for transmitting video, or limiting traffic sent by a specific computer or application to avoid congestion.

With IPv6, a flow can be one or multiple TCP connections, and a single application could generate a single flow or multiple flows. An example of a single flow would be a text page, and an example of a multiple flow would be an audio/visual conference.

Packets that share a flow label also share path, resource allocation, discard requirements, accounting, and security attributes. The flow label is defined before transmission.

Using Virtual LANs with IPv6

Virtual LAN (VLAN) is an integral feature of switched LANs. To understand and define VLANs, think of them as a group of workstations on multiple LAN segments communicating with each other as if they were on a single LAN. An example is Microsoft's technical support. Part of Microsoft's technical support is outsourced to other companies all over the country. For security and confidentiality issues, these companies' technicians are part of individual domains, trusted by Microsoft's primary domain so they can access the technical support database at Microsoft. Although based in different cities all over the world, these 'foreign' domains and users all become part on a single network.

We should take a look at the kinds of LANs available and how VLANs and IPv6 can bring value-added benefits to the corporation.

Router-based LANs

As router-based LANs are being replaced by switched LANs, VLANs are becoming an important network management tool; they are being deployed for traffic and bandwidth management in layer 2 switched networks.

As shown in Figure 1.16, router-based LANs suffer from bandwidth and latency problems, and are replaced by switched LANs. But simple switches leave a vacuum regarding network control and traffic management functionality, which is necessary to operate our networks. VLANs provide efficient tools for controlling traffic and network management; they are becoming important components and solutions for today's complex and faster networks.

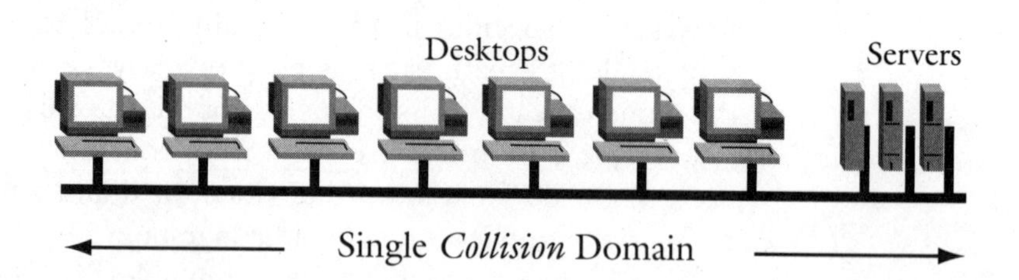

FIGURE 1.16 Typical problem faced by real LANs: collisions

Typical problems with router-based LANs are few. On a 802-type LAN, for example, a shared medium network always requires all nodes to share the bandwidth of the physical link, limiting effective utilization of the physical link. It is common for a Ethernet-based LAN to achieve only 30%-40% efficiency, since all nodes were in a single collision domain.

Switched-based LANs

In a switching-based architecture (Packet Buffering Memory), a switched port is dedicated to every node on the LAN. Each node has its own wire and all bandwidth is dedicated to the node, which eliminates the need for sharing it with all other nodes on the same LAN, as with the router-based model. Since each node is in its own collision domain, as shown in Figure 1.17, Ethernet can achieve more than 95% efficiency, promoting an almost collision-free LAN.

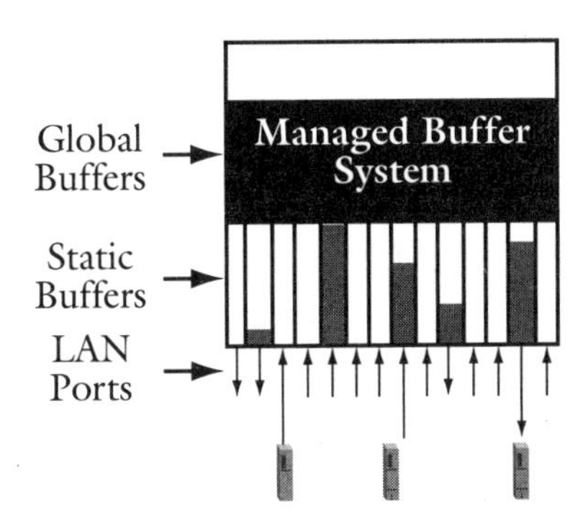

FIGURE 1.17 Switched-based LANs promote an almost collision-free network

This Ethernet efficiency is possible because a usable bandwidth within a switched LAN is determined. A switched port is only allowed to receive:

- ✔ Unicast traffic addressed to the node
- ✔ Broadcast traffic within the LAN
- ✔ Multicast traffic within the LAN.

Therefore, the volume of unnecessary broadcasts and multicasts the node receives limits this usable bandwidth to a node in a switched port.

Understanding Virtual LANs

VLANs, as we noted in the Microsoft example above, are a flexible, location/topology-independent group of stations communicating as if on a common physical LAN, as shown in Figure 1.18.

The network fabric makes sure that all nodes within the VLAN are communicating within a common broadcast domain transparently without the node's awareness. Figure 1.19 shows the list of components of a VLAN.

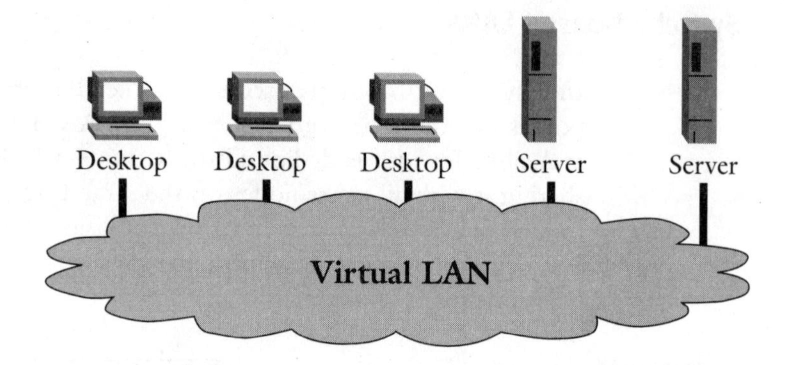

FIGURE 1.18 A VLAN connects LANs and nodes independently of their location and/or topology

FIGURE 1.19 VLAN components

Problems with a "Real" LAN

802 LAN

- A shared medium network requires all nodes to *share the bandwidth* of the physical link.
- A shared-medium network limits effective utilization of the physical link. Ethernets traditionally achieved only 30–40% efficiency because all nodes were in a *single collision domain.*

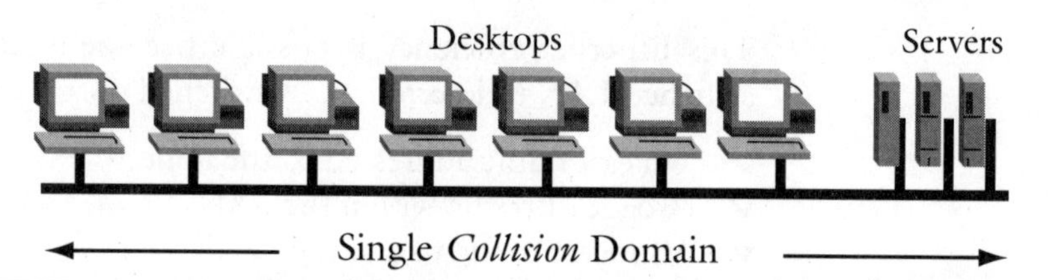

VLANs provide a series of benefits to a switched environment, including:

- ✔ Retreating from the typical Ethernet LANs, characterized by a single collision and broadcast domain
- ✔ Resolving the problem of eliminating one broadcast domain, which bridges were not able to resolve when breaking up Ethernet LANs into multiple collision domains
- ✔ Allowing for an inexpensive and fast solution for the job of routers in breaking up Ethernet into multiple collision domains and containing broadcasts within each domain
- ✔ Enabling switches to break Ethernet into multiple collision domains and use VLANs to contain broadcasts within each domain in a fast, cheap and simple way.

The VLAN bridge, which implements Virtual LAN, provides the following benefits: broadcast containment, security, and easy administration.

When VLANs are used for broadcast containment, as depicted by Figure 1.20, broadcast domain sizes can be limited to any of the following:

- ✔ IP-based, up to 1000 nodes per VLAN
- ✔ IPX-based, up to 500 nodes per VLAN
- ✔ AppleTalk-based, up to 200 nodes per VLAN
- ✔ Isolated chatty protocols.

FIGURE 1.20 Using VLAN to contain broadcasts

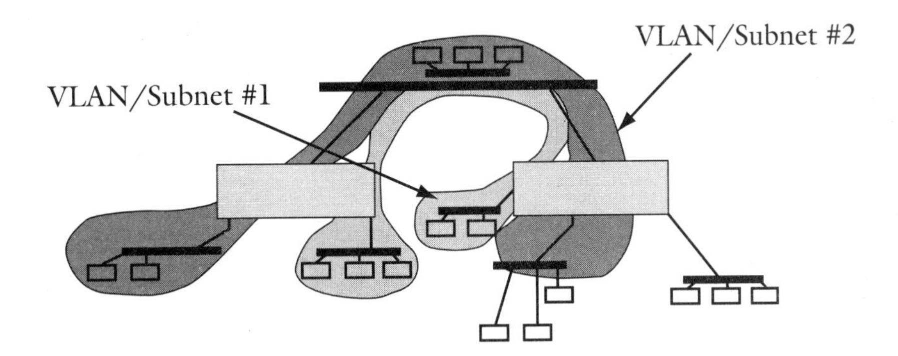

IP multicast traffic, as shown in Figure 1.21, can easily flood switched networks and VLANs can be a the best, if not the only effective, solution for resolving the ever-increasing demand on multicasting applications such as: video training, video conferencing, stock ticker and news feeds, and medical critical care, etc.

FIGURE 1.21 VLAN addresses the ever-increasing traffic of IP multicasting

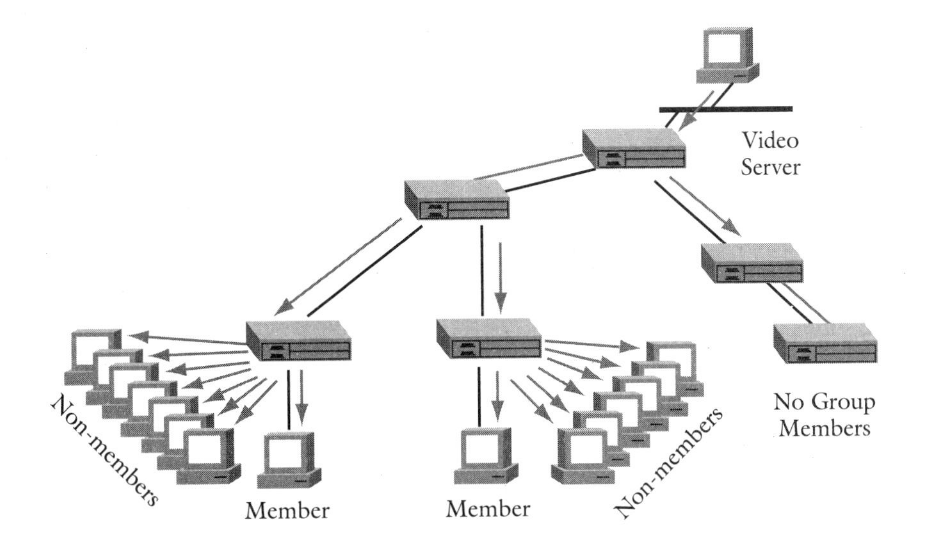

VLANs can also resolve uncontrolled proliferating IP multicasting traffic, as shown in Figure 1.22, containing it by:

- ✔ Having switches using IGMP to snoop and determine which systems want to see a multicast
- ✔ Automatically creating a autocast VLAN based upon the IGMP snooping
- ✔ Allowing multicast to go only to ports that joined the multicast group/VLAN.

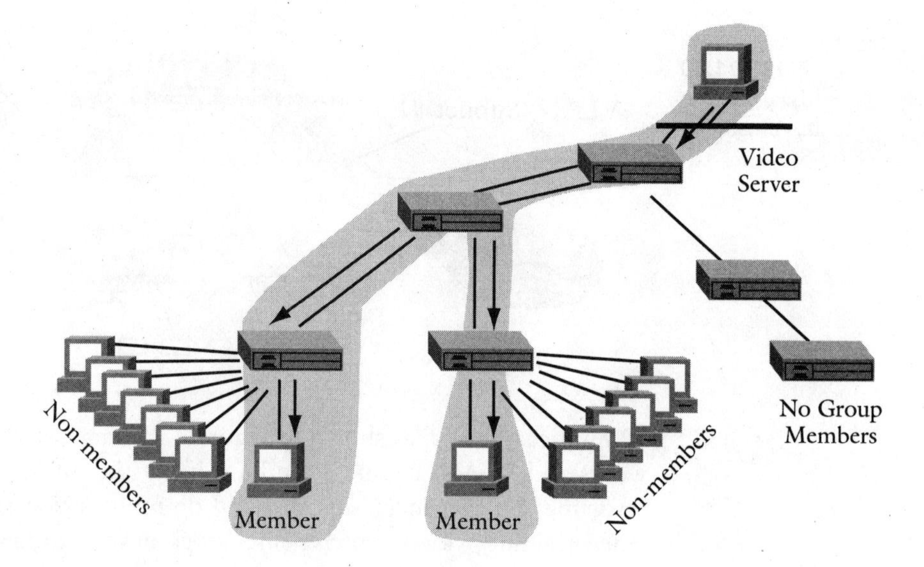

FIGURE 1.22 VLAN can be effectively used to contain the proliferation of IP multicasting

The many different kinds of virtual LANs include:

- ✔ Port-based VLAN, shown in Figure 1.23
- ✔ MAC address-based VLAN, shown in Figure 1.24
- ✔ Protocol-based VLAN, shown in Figures 1.25 and 1.26
- ✔ IP subnet-based VLAN
- ✔ IP multicast-based VLAN
- ✔ ELAN-based VLAN
- ✔ Policy-based VLAN.

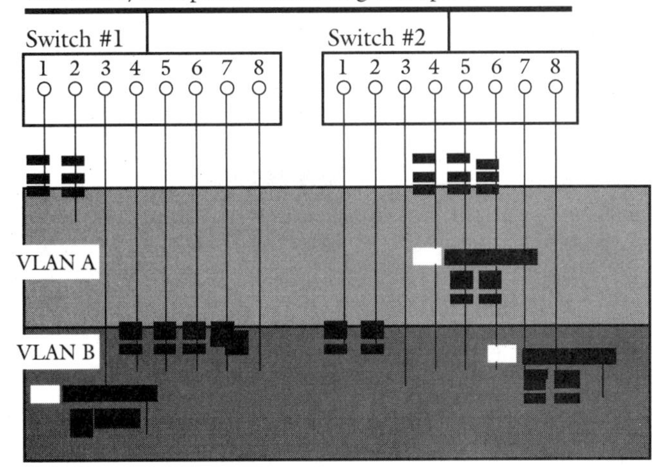

FIGURE 1.23 A typical example of port-based VLANs; Source: 3Com, Steve Jumonville presentation

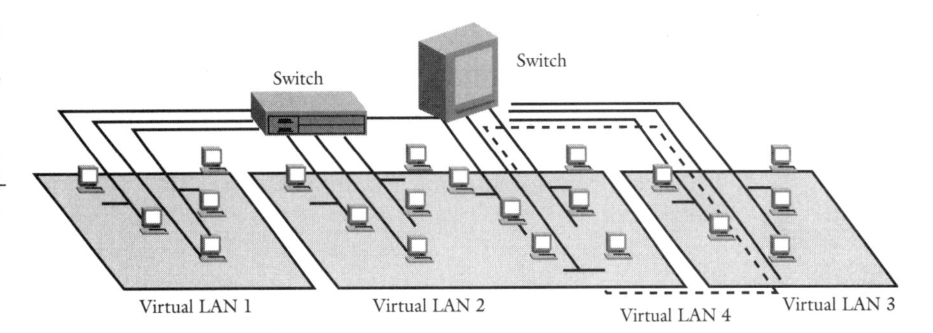

FIGURE 1.24 A typical example of MAC address-based VLANs; Source: 3Com, Steve Jumonville presentation

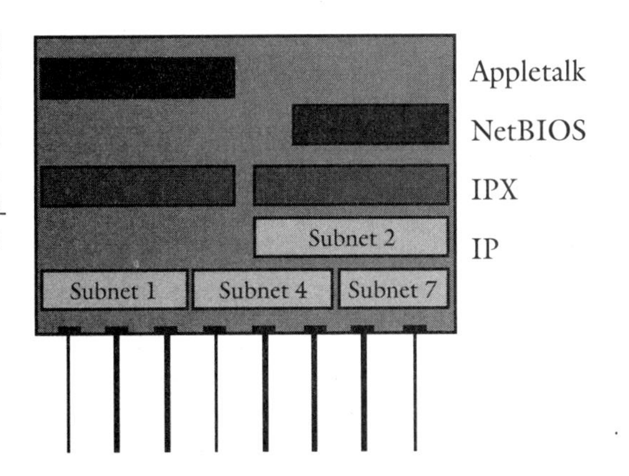

FIGURE 1.25 A typical example of Protocol-based VLANs; Source: 3Com, Steve Jumonville presentation

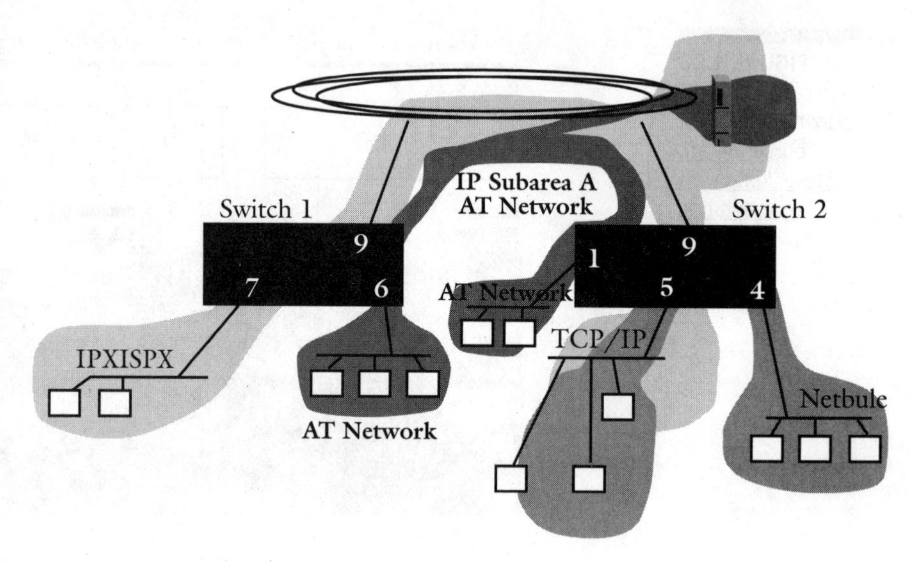

FIGURE 1.26 A protocol-based VLAN enables different protocols to be grouped together, eliminating the circulation to all machines of broadcast packets not addressed to other protocols

What's Next

This chapter discussed some of the main limitations of IPv4 and made a brief comparison of IPv4 and IPv6 addressing. The next chapter, provides the fundamentals of RSVP, IP Multicast and an overview of ATM, important technologies in carrying voice over IP.

CHAPTER 2

Understanding RSVP, IP Multicasting, and ATMs

The Resource ReserVation Protocol (RSVP) is the reservation protocol of choice on the Internet. Multicast applications, such as high-speed video transmission, which will definitely depend on a protocol like RSVP in order to guarantee high levels of "Quality of Service" (QoS), voice over IP (VoIP) and many other multimedia applications, all rely on RSVP. Many of these new applications require a different approach to routing and resource allocation than do generic data applications. RSVP is different from many IP protocols because it is receiver-driven: it is up to the receiver to select which source to receive as well as the amount of bandwidth to be reserved or paid for, assuming a commercial network.

Following the trend, QoS requirements have increased considerably with the advent of multimedia applications; multimedia applications require reliable and fast transmissions, otherwise, recipients may get "chopped" and delayed packets, because of the way lower-level protocols, such as Ethernet, ATM and token ring, handle packets to IP: the packets are forwarded to their destinations, and are therefore prone to delays or "bursts" in delivery. This can happen to all routed IP networks, because every router in the data path examines each packet of information. Usually there will be several intermediate routers between the source and the destination, joining the different networks. As the packet "hops" from one router to another, the IP protocol in each router decides which is the fastest path for that packet to go; this can easily delay the arrival of a packet at its destination, and many times the packet will never make it.

For most data applications, this "bursty" delivery is acceptable; it lends itself to high performance and high availability. For multimedia applications, involving both voice and/or video, the traffic must be "streamed" or transmitted continuously, and not in bursts. The challenge for the networking community is to accommodate these different requirements while continuing to maintain high network performance and availability.

It was to meet this challenge that IETF developed the RSVP, an end-to-end protocol compatible with current TCP/IP based networks. It is capable of providing the means to support a special QoS for multimedia applications and others that need this, while maintaining current internetworking methods, thereby preserving the existing network infrastructures.

Understanding RSVP

The RSVP protocol operates on top of IP, in the transport layer. It is a control protocol comparable to ICMP (Internet Control Message Protocol) or IGMP (Internet Gateway Message Protocol) designed to operate with the current and future unicast and multicast routing protocols. Some applications are suited for one receiver, while it is desirable with other applications to have the potential to send to more than one receiver without having to broadcast to the entire network.

The components of RSVP are:

- ✔ **Sender**—responsible for letting the receiver know there is data to be sent and what QoS is needed
- ✔ **Receiver**—responsible for sending out notices to hosts or routers so they can prepare for the upcoming data
- ✔ **Hosts or Routers**—responsible for setting aside all the proper resources.

Once all the above steps are completed, the sender can successfully send the data.

RSVP, if completely implemented, is intended to provide QoS over any media, even if the media provide none. However, RSVP allows only a much less granular, more generic QoS guarantee. The present definition of RSVP includes a number of stations, all connected to a switch that handles local traffic, which in turn is connected to a router, which provides WAN access. The RSVP definition is concerned primarily with the router, which means that when one router wants to talk to another, RSVP can request a certain quantity of bandwidth. But the Internet allows connections over various types of routers, most of which do not support RSVP.

Furthermore, RSVP does not yet apply to the station or the switch, meaning that an Ethernet card knows nothing about QoS, or limiting packet release to assure that it does not go over its allocation. These things have to be done in software via packet schedulers. There is no plan or infrastructure for putting RSVP on the switch. With ATM, every component in the line is ATM-based, so ATM can provide absolute guarantees. In addition, RSVP provides no mechanism for tracking and billing for QoS, which is a major concern for carriers. Tracking and billing are easily done using ATM.

With RSVP, the application is able to provide advance notification about the network resources it will need. By granting the reservation, the affected hosts and routers commit to providing these resources. If the router is not capable of providing them, or the resources are not available, the host or router can refuse the reservation. The application is notified right away that the network cannot support it, thereby avoiding the time and cost of a trial-and-error approach.

The two main concepts of the RSVP protocol are:

- ✔ **Flows**—These are characterized by the traffic streams from a sender to one or more receivers, and are identified in the IPv6 header by a "flow label." Prior to sending out a flow, the sender transmits a "path message" destined for the receiver. The message contains the source IP address, destination IP address and a flowspec. The flowspec, made up of the rate and delay bounds for the flow, is the QoS that the flow requires. The path message is routed to the receiver by the hosts and routers along the flow's path.
- ✔ **Reservations**—The receiver is provided with the path message and is then responsible for making the actual reservation. With the receiver making the reservation, there is greater flexibility in handling multicast flows. This receiver-based model allows for a distributed solution enabling heterogeneous receivers to make reservations tailored to their needs.

Receiver-based protocol is more effective for heterogeneous networking environments. Also, in order to assure reservations are still in place and that any "moves or changes" on the network are aware of the reservation, RSVP incorporates an approach called "soft state". This term is used because RSVP paths and reservations are considered tentative. Resources are put aside when a router accepts a reservation, but if a flow is not received, it will time out and free up its resources. With the soft state approach, the sender periodically sends its path message and the receiver continues to send its reservation request in order to refresh any time-outs or changes that may have occurred.

In summary, the list of RSVP features includes:

- ✔ RSVP makes resource reservations for unicast and multicast receivers using a soft-state approach for keeping the reservations up to date
- ✔ RSVP is unidirectional
- ✔ Receivers in RSVP initiate and maintain the resource reservation for a flow
- ✔ RSVP is not a routing protocol, but relies upon routing protocols for delivering flows
- ✔ RSVP supports IPV4 and IPV6.

The RSVP area is developing very quickly, as are efforts to construct standards. Cisco is at the moment the major player promoting RSVP, working very closely with the IETF to resolve known limitations. RSVP will certainly become a requirement for multimedia applications and router manufacturers in 1998.

IP Multicasting

Traditional Internet applications usually operate between a sender and a receiver. Emerging technologies enable a sender to communicate to a group of receivers simultaneously, as with audio/visual broadcast messages sent to a group of users, live transmission of multimedia training, transmission over networks of live TV or radio news, and so forth.

Unfortunately, such applications very often generate bottlenecks on the network, as traffic and network overhead increase tremendously, as illustrated in Figure 2.1. The proposed solution for this problem has been IP multicast, an extension of the IP standard-based solution with broad industry support. IP multicast, which has been under development since the early '90s, represents an important advance in IP networking.

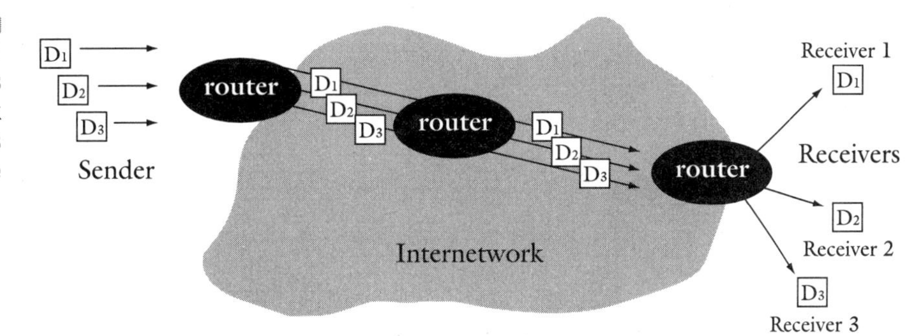

FIGURE 2.1 Broadcast messages sent over a network generate bottlenecks and poor performance

In the IP multicast scheme, each multicast group can be identified by a particular Class-D IP address, as shown in Table 2.1. Each host can register itself as a member of selected multicast groups through use of the IGMP. Thus, a whole group of recipients, members of this multicast group session, are able to receive the message, which is only broadcast to the members of the group. Only this group will benefit from the message and be affected by the traffic it generates on the network.

The remaining users on the network will not even notice it as IP multicast technologies address the needed mechanisms at different levels in the network and internetworking infrastructure for efficient handling of group communications. This scenario is shown in Figure 2.2, where only one copy of the same data (D) is "multicast" to Receivers 1, 2, and 3 in a shared conferencing application. Comparison of Figures 2.1 with 2.2 will indicate bandwidth savings, both locally and across the networks.

The set of hosts listening to a particular IP multicast address is called a host group. This group can span multiple networks. Membership in a group is dynamic—hosts may join and leave host groups, as shown in Table 2.1

TABLE 2.1 List of common multicast addresses

Well-Known Class D Address	Purpose
224.0.0.1	All hosts on a subnet
224.0.0.2	All routers on a subnet
224.0.0.4	All DVMRP routers
224.0.0.5	All MOSPF routers
224.0.0.9	RIP Version 2
224.0.1.1	Network Time Protocol (NTP)
224.0.1.2	SGI Dogfight

continued on next page

Well-Known Class D Address	Purpose
224.0.1.7	Audio news
224.0.1.11	IETF audio
224.0.1.12	IETF video

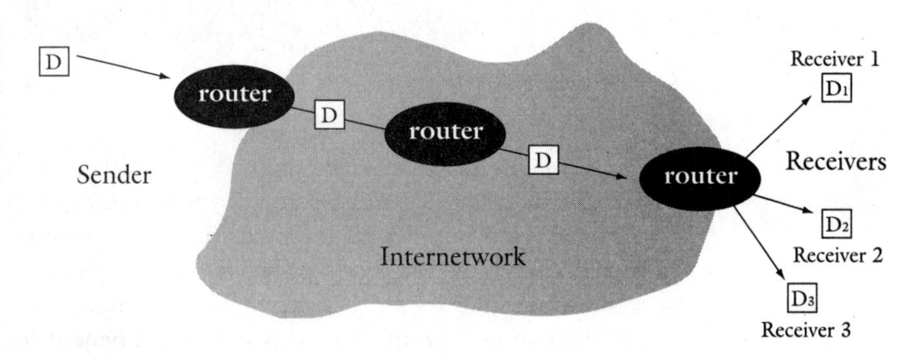

FIGURE 2.2 IP multicasting only delivers its contents to members of a multicast session

IP Multicast Benefits

One of the immediate benefits of IP multicast is the cost saving in network and server resources. The technology also enables a large number of new applications that were not available for unicasting transport.

Also, with IP multicasting, network administrators can manage network growth and control costs much more efficiently, as IP multicast is much more cost-effective than the alternatives usually deployed for increasing LAN and WAN capabilities and bandwidth. IP multicast can readily help increment network response time, as it can immediately alleviate network congestion caused by existing applications are inefficiently transmitting to groups of recipients; it can take advantage of the many applications already in place.

Another attribute of IP multicast is its scalabiltiy; it can be scaled as the number of participants and collaborations expand. Adding one or more users will not amount to adding a corresponding amount of bandwidth. Multicasting also results in a greatly reduced load on the sending server, which no longer has to support many sequential or concurrent unicast sessions, and it is fully compatible with new IP protocols and services, such as QoS requests to support real-time multimedia.

Using IP Multicast

According to Stardust Forums (**http://www.stardust.com**), which manage the IP multicast initiative, Intel Corp. deployed IP multicast on a 4,000-node Oregon site in early 1996. Intel employees regularly use IP multicast conferencing software to follow events such as conferences or executive presentations and product launches from their desktops. Toys R Us Inc. is a company using IP multicast file transfer software to send software updates to 900 store locations. Before using IP multicast, the files had to be sent over aVSAT (very small aperture terminal) nationwide network one file at a time. Because this used up so much bandwidth, it had to be performed at night. The IP multicast-based software is designed to improve product availability in the stores. A Toys R Us representative believes the system paid for itself immediately.

IP multicast technology is very important for the growth of voice over IP (VoIP) applications, as the demand for audio, video and data streams over a network is growing fast. Applications such as desktop video and audio conferencing, collaborative engineering, shared white boards, transmission of university lectures to a remote audience, and animated simulations are becoming necessities in some industries and environments. In the near future, the transmission of a corporate presentation via computer to thousands of workers seated at their desks will become reality, through the use of IP multicast. Unicast transmission would never be able to perform such a task.

Stored data streams, for updates of kiosks and Web caches, video server-to-video server updates, corporate announcements to employees, etc., will also depend upon IP multicast to be successfully deployed and delivered.

Multicast routers to keep track of group membership on each of the router's physically attached networks use IGMP messages. The following rules apply:

- ✔ A host sends an IGMP report when the first process joins a group. If multiple processes on a given host join the same group, only one report is sent, the first time a process joins the group. This report is sent out on the same interface on which the process joined the group.
- ✔ A host does not send a report when processes leave a group, even when the last process leaves a group. The host knows that there are no members in a given group, so when it receives the next query, it will not report the group.
- ✔ A multicast router sends an IGMP query at regular intervals to see if any hosts still have processes belonging to any groups. The router must send one query out on each interface. The group address in the

query is 0 since the router expects one response from a host for every group that contains one or more members on the host.

✔ A host responds to an IGMP query by sending one IGMP report for each group that still contains at least one process.

Using these queries and reports, the multicast router keeps a table of which of its interfaces have one or more hosts in a multicast group. When the router receives a multicast datagram to forward, it forwards the datagram (using the corresponding multicast link-layer address) only to the interfaces that still have hosts with processes belonging to that group.

For additional and more in-depth information about IP multicasting, refer to my book co-authored with Kitty Niles, IP Multicasting: Concepts and Applications, (McGraw-Hill).

An Overview of Asynchronous Transfer Mode Technology

Asynchronous Transfer Mode (ATM) technology is playing a major role in the development of workgroup applications, as well as university and enterprise networks, by providing scalable bandwidths, higher performance and QoS to LAN and WAN networks, enabling multimedia applications, such as voice and video over IP. Figure 2.3 describes the fundamentals of ATM.

FIGURE 2.3 ATM fundamentals

First Things First

- ATM is the transport protocol for the broadband integrated services digital network (B-ISDN)
- An ATM network is a hybrid of circuit-switched and packet-switched networks
- B-ISDN and ATM are being defined by the international digital communications community

ATM is the transport protocol selected for the broadband integrated services digital network (B-ISDN). It combines characteristics of circuit-switched networks (e.g., the telephone network) and packet-switched networks (e.g., the Internet) to yield a network capable of supporting digital data, voice, and video for synchronous and asynchronous delivery. International standards bodies, technology groups, and fora are defining B-ISDN and ATM. Contributors to the definition of B-ISDN and ATM include equipment manufacturers, service providers, government agencies, universities, and user entities.

The ATM Forum is generally perceived as the core technical body with respect to ATM standards and technical definition, as outlined in Figure 2.4. It is more appropriate to define the ATM Forum as the body which coordinates the interoperability and adoption of ATM between existing standards bodies and fora. In fact the ATM Forum does *not* produce ATM standards—the ITU-T is the organization which produces international standards (recommendations), as depicted in Figure 2.5. The ATM Forum works closely with groups such as the Frame Relay Forum and the Digital Audio Video Council, to promote interoperation with accepted and emerging digital communications technologies.

FIGURE 2.4 ATM standard bodies and fora

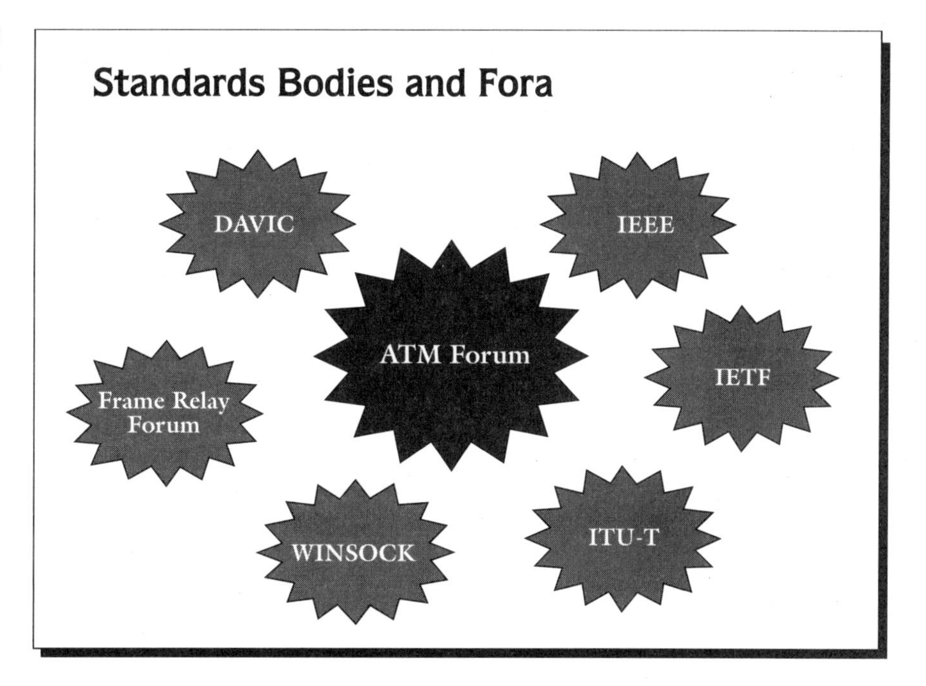

Since the ATM Forum is in place to accelerate the implementation and acceptance of ATM, it does not generate international standards. Standards related to broadband ISDN and ATM (B-ISDN's transport method) are defined in the international telecommunications union—namely, the ITU-T group (formerly the CCITT). Digital transmission standards are included in the G-series recommendations—SDH specifications are included here (SONET is the ANSI standard found in the U.S.). ISDN and B-ISDN standards are included in the I-series recommendations—these include the ATM and AAL specifications. The Q-series recommendations comprise ISDN and B-ISDN signaling standards.

FIGURE 2.5 The ITU-T standards

The ITU-T Standards

Signaling Q.2931
Traffic Mgmt I.371
ATM Adaptation Layer I.362 I.363 I.364
ATM Layer I.361
SONET/SDH 6.707 6.708 6.709

The recently overhauled ATM Forum Web site, as shown in Figure 2.6, gives all interested in ATM the opportunity to track progress of ATM standards and activities. In addition, all completed technical specifications may be downloaded from this site—free of charge. Point your browser to **http://www.atmforum.com**.

The current technical working groups in the ATM Forum are listed in Figure 2.7. The status of specification development in these groups is included at the ATM Forum Web site under technical specifications.

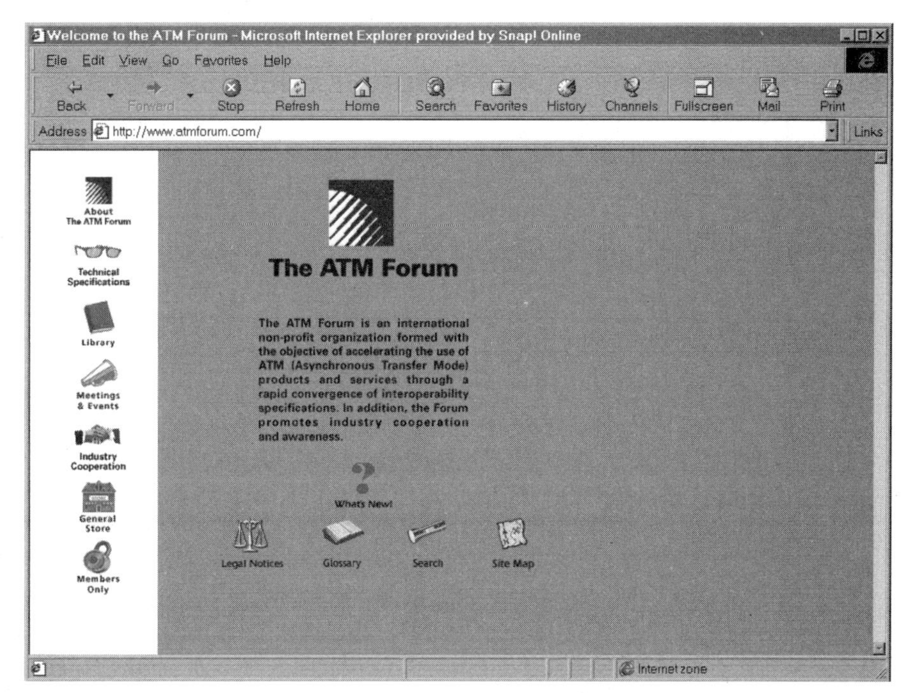

FIGURE 2.6
The ATM Forum Web site

FIGURE 2.7
Current technical working groups in the ATM Forum

Technical Working Groups

- Broadband Inter-Carrier Interface [B-ICI]
- LAN Emulation and multiprotocol over ATM [LANE/MPOA]
- Private Network-to-Network Interface [P-NNI]
- Signaling
- Traffic Management
- Voice and Telephony over ATM [VTOA]
- Wireless ATM [W-ATM]
- Network Management
- Physical Layer
- Security
- Testing
- Service Aspects and Applications [SAA]
- Residential Broadband [RBB]
- Joint PHY and RBB

ATM Technical Specifications

The ATM Forum technical working groups create technical specifications in the general areas specified in Figure 2.8. Physical layer specifications include 25.6 Mbps over UTP, 155 Mbps over multimode fiber, 51 Mbps over plastic optical fiber, and others. Management specifications address network management issues for private and public networks, and public-private network interfaces.

FIGURE 2.8 ATM technical specifications

ATM Technical Specifications

- Physical Layer
- Testing
- Management
- Private and public network interface
  - UNI, PNNI, B-ICI
- Services
  - LANE/MPOA, VTOA, ATM APIs, MPEG2 support

Private and public network interface specifications include the user to network interface (UNI), the private network-to-network interface, and the broadband intercarrier interface. ATM service specifications include LAN emulation, multiprotocol over ATM, voice telephony over ATM, ATM application programming interfaces, and others.

ATM is compatible with existing physical networks such as twisted pair, coax and fiber optics, because it isn't design-limited to a specific type of physical transport. To its advantage, unlike conventional LANs, ATM has no inherent speed limit. In contrast, when Ethernet speed was increased from ten to 100 megabits per second, its architecture required a reduction in the length of Ethernet segments from 2,500 meters to 250 meters. Similarly, Token Ring has gating factors on its speed. With ATM, there is nothing in the architecture that limits speed; an ATM network can operate as fast as a physical layer can be made to run.

Furthermore, while 100-megabit Ethernet and other high-speed networks can provide comparable bandwidth, only ATM can provide the QoS guarantees required for confidently deploying real-time telephony, video streaming, smooth videoconferencing, and other no-delay voice and video applications. QoS is so vital to the deployment of multimedia applications over IP that a number of initiatives are under way to provide QoS support for TCP/IP based networks, including the RSVP protocol specification discussed earlier in this chapter.

Private and public interface specifications include public and private UNI, P-NNI, and B-ICI, as shown in Figure 2.9. The user-to-network (UNI) document specifies how a user maintains connections to a private of public ATM network. The private network-to-network (P-NNI) document specifies how a private ATM network node signals connections through the network and maintains a routing hierarchy between ATM network nodes. The broadband intercarrier interface (B-ICI) document specifies how public network edge nodes signal and share information with other public networks.

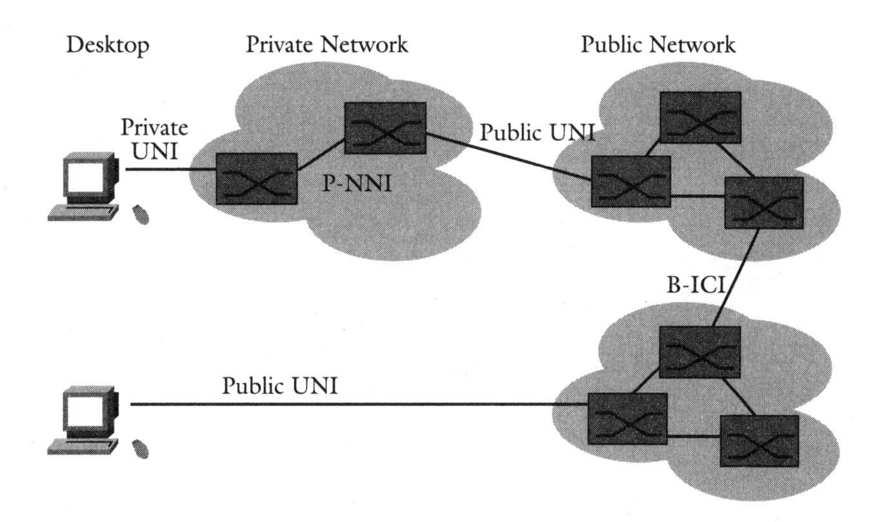

FIGURE 2.9 ATM's interface specifications

ATM's Mission: The Delivery of Multimedia Services

ATM is simply the transport mechanism chosen for the broadband ISDN network. The goal is to offer data, voice, and video services. Services imply end-to-end application connections, which transfer digital information, namely IP data, voice, and video, as depicted in Figure 2.10.

For example, the LAN emulation and multiprotocol over ATM services enable transfer of data traffic over an ATM infrastructure. Voice and telephony over ATM enable use of an ATM infrastructure for transmitting real-time voice. The MPEG over ATM function allows for transfer of encoded digital video over the ATM network.

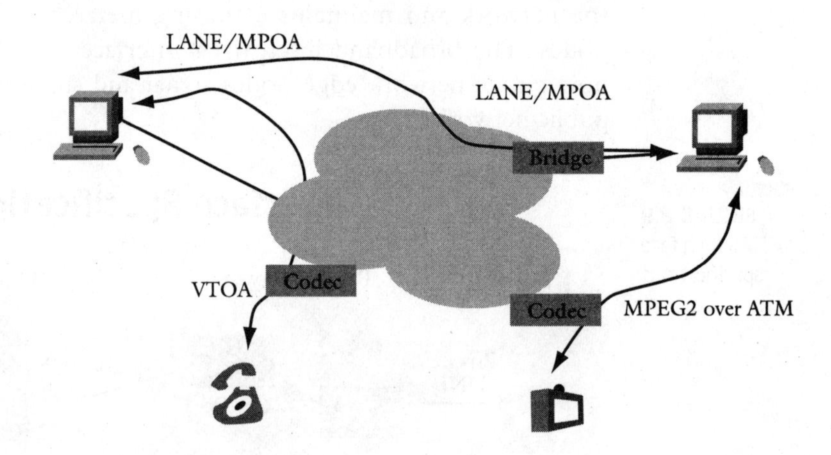

FIGURE 2.10
ATM's service specifications

Have you heard about the Anchorage Accord? This is an attempt to ease fears that ATM specifications and addenda were preventing vendors from reasonably building interoperable, and backward-compatible ATM solutions, as shown in Figure 2.11. To that end, The ATM Forum penned an agreement, including a large set of specifications, which together constitute the foundation guidelines for building an ATM system.

ATM Network Design and Solutions

ATM network design and solutions require that the following important components be addressed (see Figure 2.11):

- ✔ Review of the ATM data model
- ✔ Mapping solutions to hardware and software
- ✔ Introducing legacy network designs
- ✔ Migration and upgrade strategies to ATM solutions.

ATM Network Client Models

Figure 2.12 provides a list of ATM network client devices. Note that this is not an exhaustive list of client devices. There are other ATM-attached devices that perform adaptation such as the cells-in-frames attachment device (CIF-AD), which works in concert with Ethernet clients to provide an ATM-over-Ethernet function, which will be discussed in more detail later in this chapter. An ATM concentrator is an example of an ATM-attached device that performs no adaptation (ATM WG switches are used in place of these devices today).

FIGURE 2.11 ATM's network design and solutions

Network Solutions and Designs

- Review ATM Data Model
- Map solutions to hardware and software
- Introduce legacy network designs
- Plan migration and upgrade strategies to ATM solutions

FIGURE 2.12 ATM network client devices

ATM Network Clients

- ATM-attached LAN devices and endstations adapt legacy application behavior to ATM network
- ATM workgroup switches are ATM-only
- Legacy-attached devices rely on a higher-layer device to do adaptation

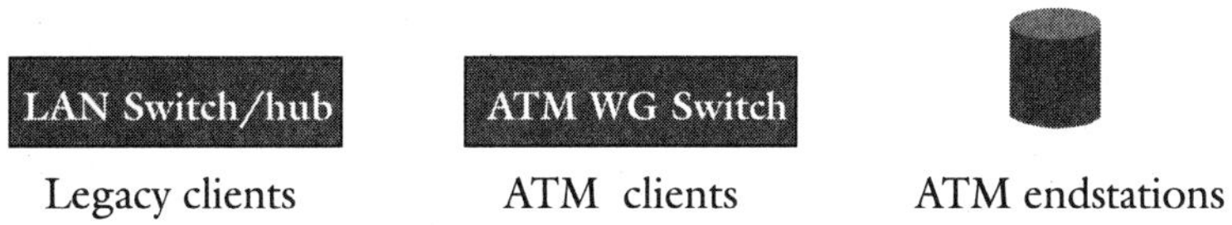

LAN Switch/hub
Legacy clients

ATM WG Switch
ATM clients

ATM endstations

Figure 2.13 indicates types of devices supporting client and server software function associated with ATM data protocols. Figure 2.14 is a rundown of the protocols supported by ATM technologies, which include but are not limited to:

- ✔ ARIS—aggregate route-based IP switching (IBM)
- ✔ ARP—address resolution protocol
- ✔ BUS—broadcast and unknown server
- ✔ CIP—classical IP
- ✔ GSMP—general switch management protocol (Ipsilon)

- ✔ IFMP—Ipsilon flow management protocol
- ✔ LEC—LAN emulation client
- ✔ LECS—LAN emulation configuration server
- ✔ LES—LAN emulation server
- ✔ MARS—multicast-address resolution server
- ✔ MPOA—multiprotocol over ATM
- ✔ NHRP—next-hop resolution protocol
- ✔ TDP—tag-distribution protocol (Cisco)
- ✔ UNI—user-to-network interface.

FIGURE 2.13 Mapping solutions to products

Mapping Solutions to Products

	CIP		LANE		MPOA		rfc1483		IP switch	
	client	srvc.	client	srvc.	client	srvc.	client	srvc.	client	srvc.
ATM BB switch	✔	✔	✔	✔	✔	✔	✗	✔*	✗	✔*
Router	✔	✔	✔	✔	✔	✔	✔	✗	✔	✔
ATM WG switch	✔	✔	✔	✔	✔	✔	✗	✗	✗	✗
LAN switch	✔	✔	✔	✔	✔	✗	✗	✗	✔	✗
Adapter	✔	✔	✔	✔	✔	✗	✗	✗	✔	✗

✔* signaling only

FIGURE 2.14 ATM's protocol support

And the Protocol Is ...

	CIP	LANE	MPOA	rfc1483	IP switch
Client Software	▪CIP client	▪LEC ▪proxy LEC	▪LEC ▪proxy LEC ▪MPOA-c ▪proxy MPOA-c	Encap. ONLY	▪IFMP-c ▪TDP-c ▪ARIS-c
ATM Network Services	▪ATM ARP ▪MARS ▪NHRP ▪UNI	▪LES ▪BUS ▪LECS ▪UNI	▪LES ▪BUS ▪MARS ▪NHRP ▪Default forwarder ▪UNI	▪UNI	▪GSMP

Note that CIP, LANE, and MPOA follow a client/server model where the ATM network provides ARP, multicast, and routing services. RFC 1483 is not a protocol, but an encapsulation method. ATM network provides only signaling (via GSMP) for IP switching clients.

ATM's VLAN Model

In the VLAN model, as depicted in Figure 2.15 and outlined in Figure 2.16, clients are members of VLANs.

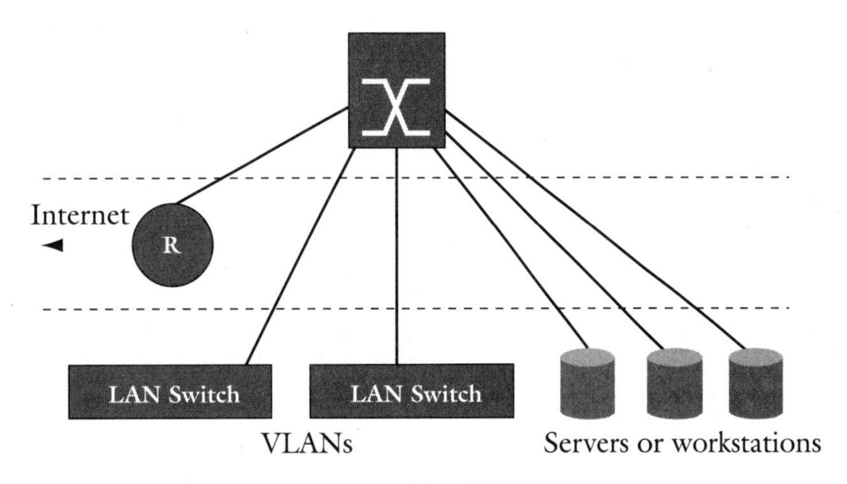

FIGURE 2.15 ATM's network model: VLAN

FIGURE 2.16 Characteristics of a VLAN model

VLAN Model

- Bridged network paradigm
- VLANs used to isolate multicast domains
- ATM LANE and MPOA inherently support dynamic VLANs
- LANE requires routing for inter-VLAN
- MPOA includes a routing service

The VLAN model, as depicted in Figure 2.16, is topologically simpler than a routed model because there are no intermediate devices between the LAN access and the backbone. VLANs allow for building non-geographically dependent LANs.

ATM LANE is an ATM-based VLAN solution. This is a good solution for networks which are currently bridged. LANE may prove to be inefficient if there is too much inter-VLAN (emulated LAN) communication—since all inter-VLAN traffic must travel through a router. A future LANE version is slated to incorporate cut-through, inter-VLAN communications. ATM MPOA is another ATM-based VLAN solution. Actually, LANE is a subset of MPOA—MPOA with no routing services is LANE. MPOA adds cut-through routing between LANE VLANs: the MPOA clients can establish connections with other clients on different emulated LANs (by learning the route from the MPOA route service).

In a VLAN model, members of the same VLAN can establish direct connections among themselves by the very nature of the LANE protocol (members of the same ELAN are associated with the same LANE servers), as shown in Figure 2.17. Communication between VLANs requires routing services, provided either by a one-armed router (LANE) or by MPOA route services. Basically, MPOA route service provides cut-through relief for LANE inter-VLAN traffic by forwarding routing information to clients. With this routing information clients are able to establish direct connections with other clients associated with different VLANs.

FIGURE 2.17 Cut-through in a VLAN model

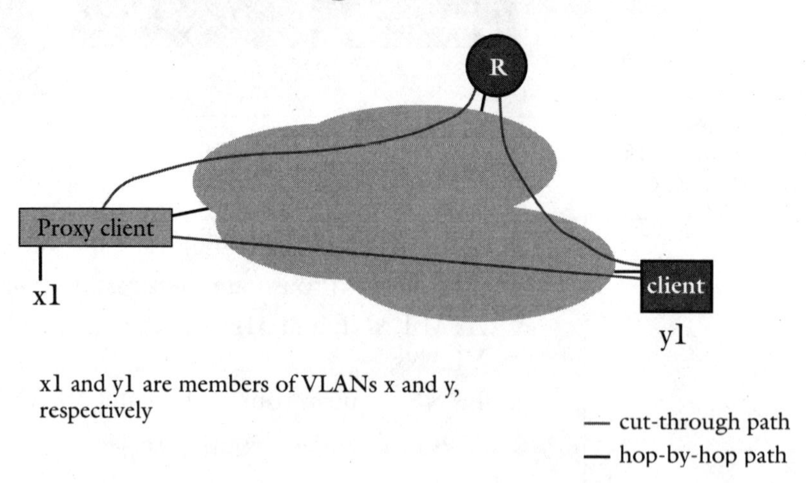

It is interesting to note that even though LAN emulation requires an one-armed router for inter-ELAN communications, if such traffic is limited, a single one-armed router is sufficient. The implication is that one could guarantee a single router hop for inter-ELAN communication—obviating the need to bother with cut-through if avoiding a single router hop for minimal traffic is the object.

ATM's Routed Model

In the routed model, as shown in Figure 2.18 and outlined in Figure 2.19, clients are members of subnets. Routed models use Layer 3 networking concepts to build multicast domains (subnets). RFC1483 encapsulation allows for adding capacity to an existing routed network by adding ATM links between routers and establishing (a mesh of) PVCs over these links.

FIGURE 2.18 ATM's routed model

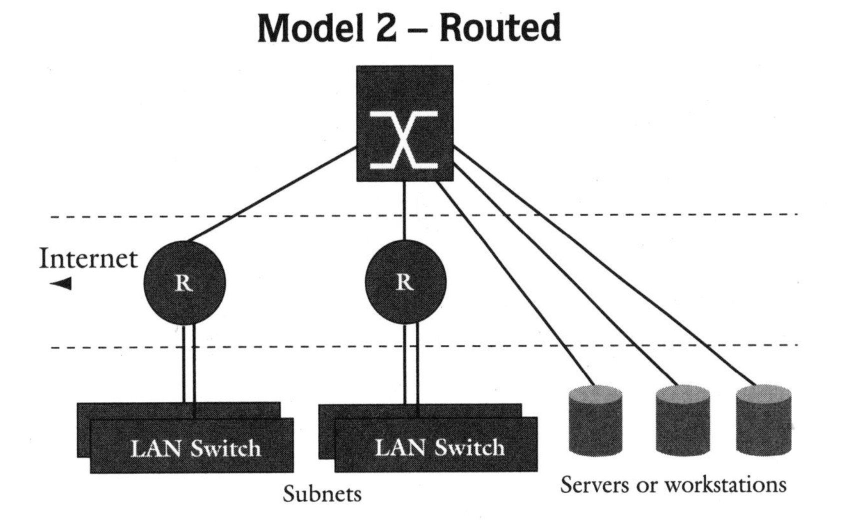

FIGURE 2.19 Characteristics of ATM's routed model

Routed Model

- Routed network paradigm
- Clients are members of subnets
- Collapsed backbone approach
- RFC 1483 adds ATM links to an existing router backbone
- IP switching/MPLS add dynamic, cut-through switching to a routed backbone

IP switching adds ATM switching to a routed backbone and also brings dynamic cut-through. Dynamic cut-through in a routed backbone means that the IP switching protocol automatically establishes a path through the ATM switching cloud and between the ingress and egress routers—avoiding hopping through intermediate routers

In Figure 2.20, a hop-by-hop routing path has been established through routing configuration (say an OSPF path). This path is associated with communications between subnet x and subnet y. IP switching (or MPLS) establishes another more direct path which bypasses intermediate routers—a cut-through path.

FIGURE 2.20 Cut-through in a routed model

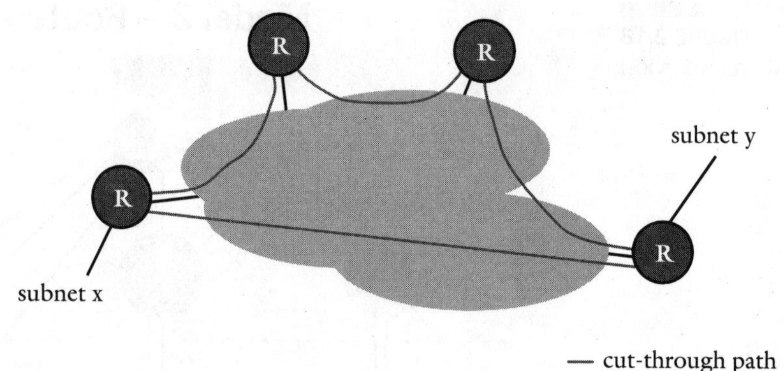

IP switching protocols differ mainly in how and when the decision is made to establish a cut-through path. Ipsilon and many partners currently ship a solution like this. Other proposed solutions include Cisco's Tag switching and IBM's Aggregate route-based IP switching (ARIS).

In summary, LANE, MPOA, and IP switching all include methods for establishing edge-to-edge connections through the ATM network. Since this is the case, the performance through the ATM network is identical regardless of the protocol (once a cut-through path is established). Thus, one can implement the edge strategy that best suits the desired network behavior, whether that be router-based or VLAN-based (and get the same benefits of cut-through). Figure 2.21 provides a summary of ATM's network service characteristics and Figure 2.22 a brief outline of ATM switch testing procedures.

FIGURE 2.21
ATM network services summary

Summary

- LANE, MPOA, RFC 1483, and IP switching enable ATM-based data networking
- RFC 1483 preserves hop-by-hop routing
- MPOA and IP switching provide mechanisms for employing routing while avoiding router hops (cut-through)
- LANE, MPOA, and IP switching can establish cut-through paths

FIGURE 2.22
ATM testing procedures

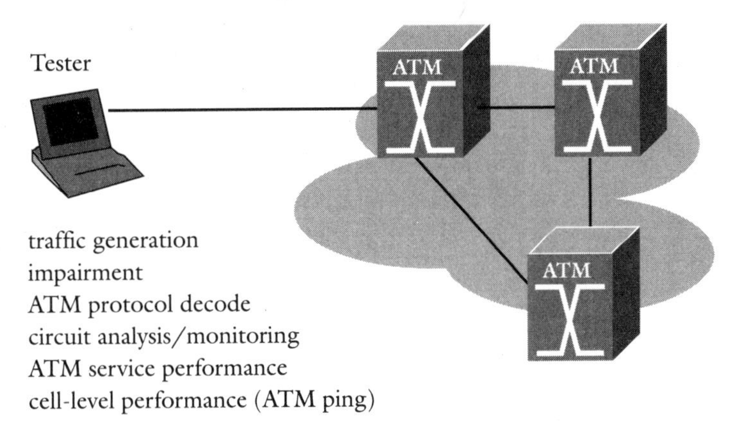

ATM is Connection-Oriented

The better to appreciate connection-oriented ATM, it is helpful to review connectionless systems. LAN architecture, whether Ethernet, token ring, or FDDI, share certain characteristics. Each station is connected to the network via an adapter card, which has a driver, above which is a protocol driver, such as TCP/IP. In traditional LANs, such as Ethernet, the driver protocol is connectionless, meaning that the protocol driver simply provides a packet with a source address and a destination address and sends it on its way. Being joined by a common medium, each station will see the packets of data put on the wire by each of the others, regardless of whether the packet is passed sequentially, as in a ring topology, or broadcast, as with

Ethernet. The primitive message from the station to the wire, or from the protocol to the adapter, is simply "send packet."

Once the packet has been sent, according to the specifications of whatever LAN is being used, the adapter knows that the packet is visible to all stations on the network. Each station has an adapter card, which processes the packet and examines the destination address. If the address applies to that machine, the adapter does a hardware interrupt and accepts the packet. If not, the adapter parses it. Again, this is called connectionless because no logical connection to the recipient address was made, the packets were simply addressed and put onto the network.

A LAN network such as Ethernet offers very few services, because all an Ethernet card can do is take a packet and send it. Being connectionless, it can provide no guarantees or similar features. For example, it cannot determine the status of the target machine. This is why developers rarely write applications directly to Ethernet. Rather, protocol drivers are used to enter sequence numbers, verify packet arrival (retransmitting, if necessary), partition big messages into smaller ones, and such—with all of these services adding time to the transmission, and with none of them able to provide end-to-end QoS guarantees.

However, ATM is complex. Even though ATM cells and cell switching make easier for the development of hardware-intensive, high-performance ATM switches, its deployment requires a very complex and intensive integration of software and protocol infrastructure. This is especially true when linking individual ATM switches into a network, as well as internetworking the network with the vast installed base of existing local and wide area networks.

TIP

ATM cells have a fixed length of 53 bytes. When fixed-length cells are used, the information can be transported in a predictable manner. This predictability accommodates different traffic types on the same network—for example, voice, data, and video.

The ATM cell is broken into two main sections, the header and the payload, as shown in Figure 2.23. The header (5 bytes) is the addressing mechanism and is significant for networking purposes as it defines how the cell is to be delivered. The payload (48 bytes) is the portion that carries the actual information—either voice, data, or video. (The payload is also referred to as the user information field.) An ATM cell is shown below.

FIGURE 2.23
ATM cell is broken into two main sections, the header and the payload

Bytes	5	48
	Header	User data

This section does not cover all the aspects of ATM technology, as there are already books available on the market on this subject. For more information, check the end of this book. I assume readers have at least a basic knowledge about ATM layer protocols and cell formats, as well as of the operation of ATM switching systems. ATM is treated here in the context of carrying IP transmissions, more specifically voice over ATM, as discussed in more detail in Chapter 3.

Basic Understanding of ATM Networks

An ATM network consists of a set of ATM switches interconnected by point-to-point ATM links or interfaces. These ATM switches support two kinds of interfaces: user-network interfaces (UNI), which connects ATMs end-systems such as hosts, routers, etc., to an ATM switch, and network-node interfaces (NNI), which are any physical or logical links across which two ATM switches exchange the NNI protocol, as shown in Figure 2.24.

FIGURE 2.24 ATM network interfaces

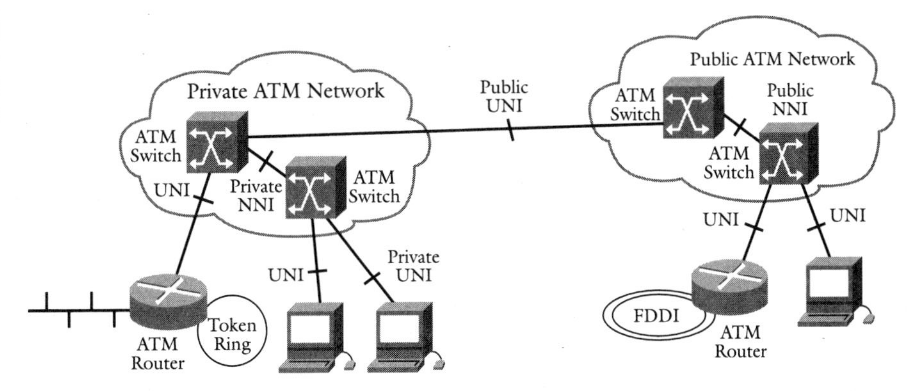

Fundamentally, ATM networks are connection-oriented, requiring virtual circuits to be set up across the ATM network before any data can be transferred. There are two types of ATM circuits:

- ✔ **Virtual paths**, identified by virtual path identifiers (VPI), are bundles of virtual channels, switched transparently across the ATM network on the basis of the common VPI.
- ✔ **Virtual channel**, identified by the combination of a VPI and a virtual channel identifier (VCI), have only local significance across a particular link, and are remapped, as appropriate, at each switch.

Although ATM switch implementations are complex, the basic operation is simple, as shown in Figure 2.25:

- ✔ The switch receives a cell across a link through a known VCI or VPI value
- ✔ It looks up the connection value in a local translation table to determine the outgoing ports of the connection and the new VPI/VCI value of the connection on that link
- ✔ It then retransmits the cell on that outgoing link with the appropriate connection identifiers.

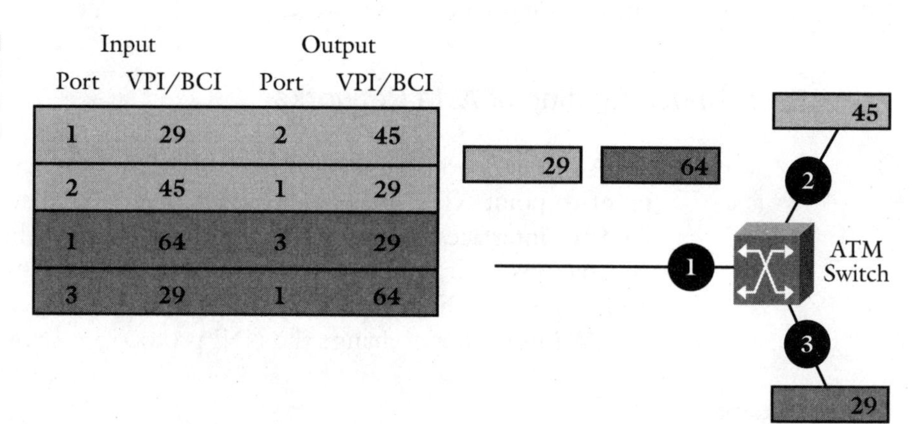

FIGURE 2.25 ATM switch operations

This simple switch operation is shown in Figure 2.26. External mechanisms are responsible for the setup of the local translation tables prior to the transmittal of any data. The manner in which these tables are set up determines the two fundamental types of ATM connections:

- ✔ **Permanent Virtual Connections (PVC)**—A PVC is a connection set up by some external mechanism, typically network management, in which a set of switches between an ATM source and destination ATM system is programmed with the appropriate VPI/VCI values.
- ✔ **Switched Virtual Connections (SVC)**—An SVC is a connection that is set up automatically through a signaling protocol. SVCs do not require the manual interaction needed to set up PVCs and are thus likely to be much more widely used. All higher-layer protocols operating over ATM primarily use SVCs, and it is these that are primarily considered in this book.

ATM signaling is initiated by an ATM end-system that desires to set up a connection through an ATM network, as shown in Figure 2.27. Signaling packets are sent on a virtual channel, VPI=0, VCI=5. The signaling is then routed through the network, from switch to switch, setting up the connection identifiers as it moves along, until it reaches the destination end-system. The latter can either accept/confirm the connection request, or reject it, causing the clearance of the connection.

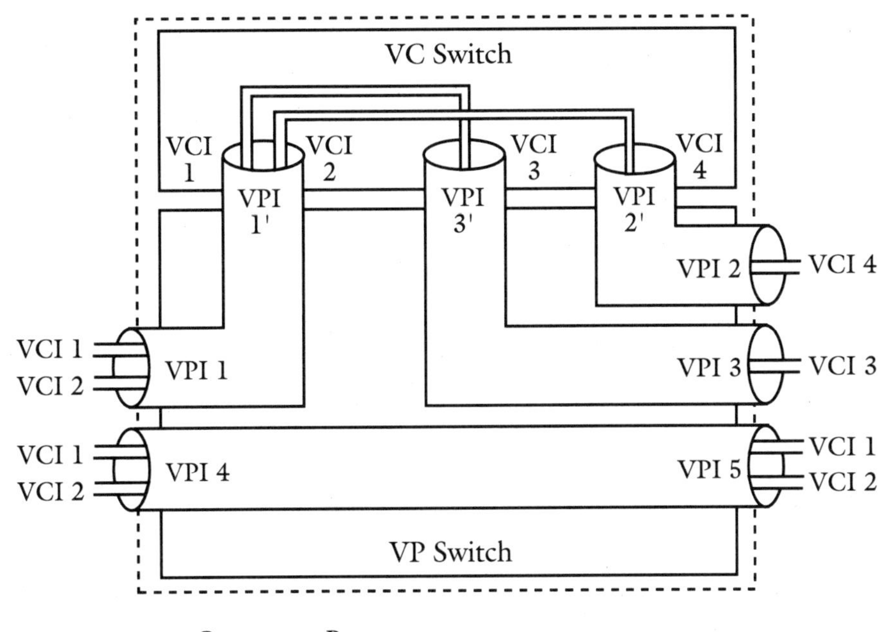

FIGURE 2.26 A diagram of ATM's two types of connections: virtual circuit and virtual patch switch

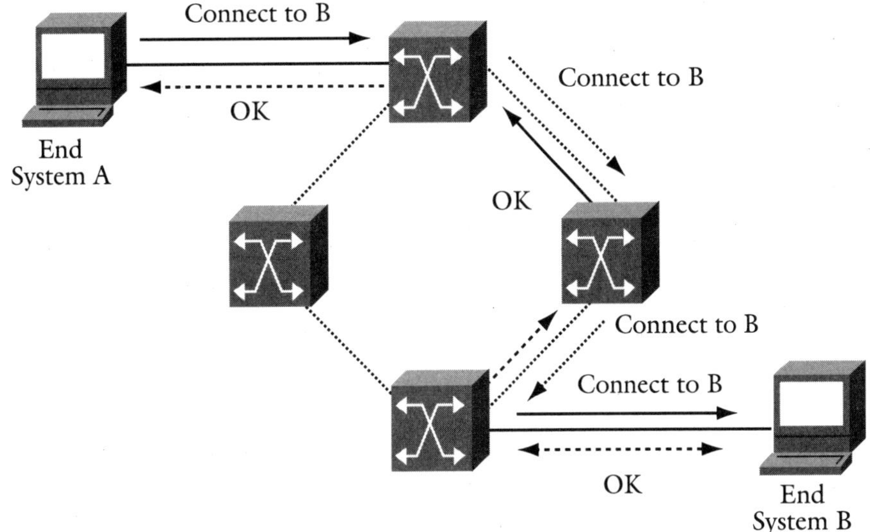

FIGURE 2.27 Setting up a connection through ATM signaling

There are two fundamental types of ATM connections, as illustrated in Figure 2.28:

- ✔ **Point-to-point connections**—these connect two ATM end-systems and can be unidirectional or bidirectional.
- ✔ **Point-to-multipoint connections**—these connect a single source end-system to multiple destination end-systems. The ATM does cell replication within the network switches at which the connection splits into two or more branches.

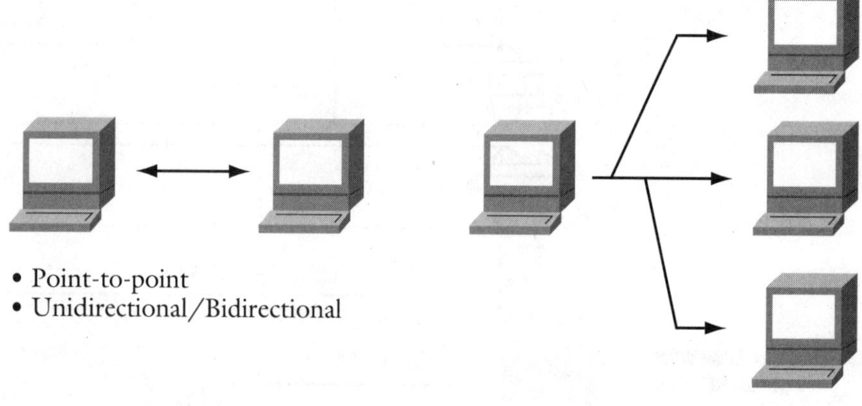

FIGURE 2.28 Types of ATM connections

These types of ATM connections do not have any analog to multicasting or broadcasting capabilities so common in shared-medium LAN technologies such as Ethernet or token ring. In such technologies, multicasting allows multiple end systems both to receive data from other multiple systems, and to transmit data to these multiple systems. Such capabilities are easy to implement in shared-media technologies such as LANs, where all nodes on a single LAN segment must necessarily process all packets sent on that segment. In an ATM-to-multicast LAN group, the only analog would be a bidirectional multipoint-to-multipoint connection. However, this alternative cannot be implemented if the ATM Adaptation Layer (AAL) 5, one of the most common, is used to transmit data across ATM networks.

Unlike AAL 3/4, with its Message Identifier (MID) field, AAL 5 does not provide for the interleaving of cells from different AAL5 packets on a single connection within its cell format. Thus, all AAL5 packets sent to a particular destination across a particular connection must be received in sequence, with no interleaving between the cells of different packets on the same connection. Otherwise, the destination reassemble process would not be able to reconstruct the packets.

This is why ATM AAL 5 point-to-multipoint connections can only be unidirectional. If a leaf node were to transmit an AAL 5 packet onto the connection, both the root node and all other leaf nodes would receive it. However, at these nodes, the packet sent by the leaf could well be interleaved with packets sent by the root, and possibly other leaf nodes; this would preclude the reassembly of any of the interleaved packets. Clearly, this is not acceptable.

Nevertheless, ATM *does* require some form of multicast capability, since most existing protocols, developed initially for LAN technologies, rely upon the existence of a low-level multicast/broadcast facility. Therefore, three methods have been proposed for solving this problem:

- ✔ **VP-Multicasting**—A multipoint-to-multipoint VP links all nodes in the multicast group, and each node is given a unique VCI value within the VP, allowing interleaved packets to be identified by the unique VCI value of the source. However, this mechanism requires a protocol for unique allocation of VCI values to nodes; this mechanism does not currently exist.
- ✔ **Multicast Server**—All nodes wishing to transmit onto a multicast group set up a point-to-point connection with an external device, known as a multicast server, as illustrated in Figure 2.29. This multicast server is then connected to all nodes wishing to receive the multicast packets through a point-to-multipoint connection, receiving the packets, confirming they are serialized, and retransmitting them across the point-to-multipoint connection

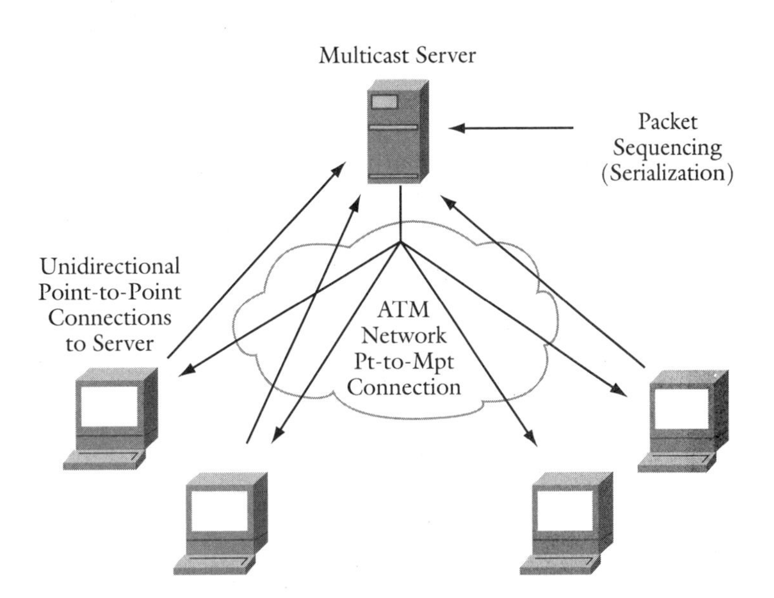

FIGURE 2.29
Typical example of a multicast server operation

✔ **Overlaid Point-to-Multipoint Connections**—All nodes in the multicast group establish a point-to-multipoint connection with every other node in the group, as shown in Figure 2.30, becoming leaves in the equivalent connections of all other nodes, enabling all of them both to transmit to and receive from all other nodes.

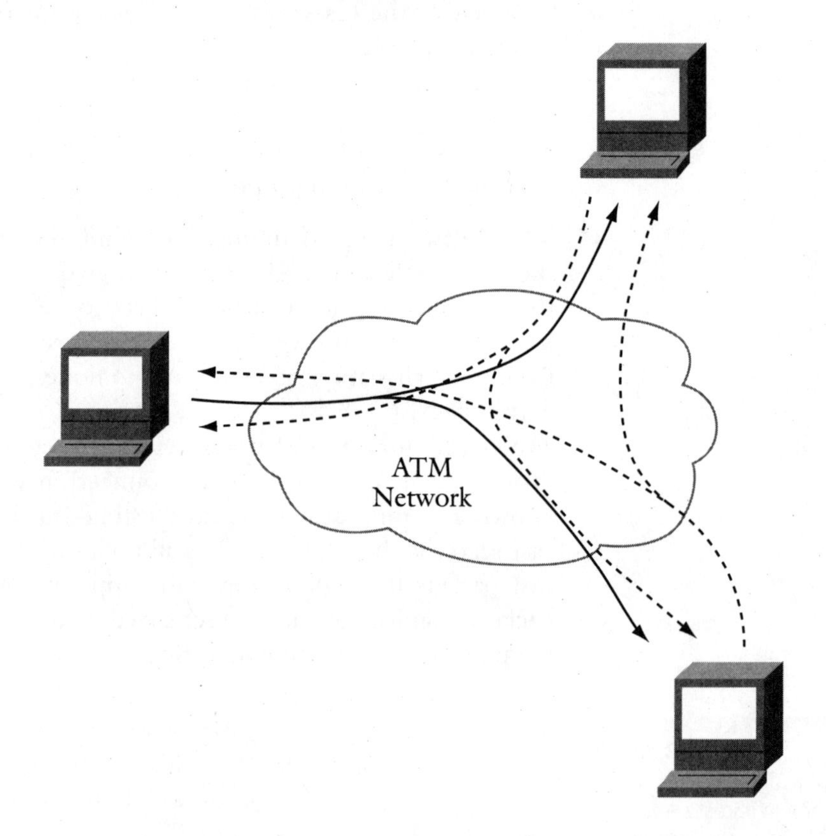

FIGURE 2.30 Multicast through overlay of point-to-multipoint connections

The overlaid point-to-multipoint mechanism requires each node to maintain N connections for each group, where N is the total number of transmitting nodes within the group, while the multicast server mechanism requires only two connections. It also requires a registration process for telling nodes that join a group what the other nodes in the group are, so that their own point-to-multipoint connection can be formed.

As we have noted above, there is no recommended solution for the use of ATM for multicast, which affects its ability to carry voice over IP successfully. Higher-layer protocols within ATM networks use both the second and third solutions for multicast, as discussed in more details in Chapter 3. This is one example of why internetworking existing protocols with ATM is so complex. Most current protocols, particularly those developed for LANs, implicitly assume a network infrastructure of a shared medium and connectionless technology with implicit broadcast mechanisms. ATM technology violates all of these assumptions!

Since this book proposes to discuss voice over IP, I decided not to cover ATMs any more than was necessary to understand how IP multicasting and VoIP rely on ATM. Further discussion would be outside the scope of this book. For additional information on ATMs, such as ATM signaling and addressing, consult Cisco's Web site, which has a variety of resources and documents on ATM and other IP technologies. Cisco's URL is **http://cio.cisco.com/**.

What's Next

This chapter introduced the fundamentals of the RSVP protocol, IP multicasting and ATM technologies, as these technologies are key to the development and deployment of voice over IP technologies and services. Chapter 3 introduces the basic concepts of voice over IP, and its most-used H.323 standard. It discusses other standards and technologies such as audio codecs, IP over ATM, voice over ATM, the emulation of traditional T1/E1 Trunks, IP over SONET and voice over SONET, and IP and voice over frame relay. It also treats Layer 3 switching and gigabit Ethernet as well as their role in VoIP.

[illegible]

[illegible]

What's Next

[illegible]

CHAPTER 3

IP Superhighway

Voice over IP

Not so long ago the Internet used to run on phone systems, but now, phone systems are running on the Internet. Voice and data consolidation can bring tremendous savings in communications expenses, up to 90 percent. If the implementation of voice over IP technology is done right, the savings can be very significant, amounting to up to 90 percent, and many times enabling a full return of investment (ROI) in less than a year. Furthermore, the savings do not stop, as the ongoing costs of maintaining voice over IP are much less those of phone systems.

Since the 1980s companies have carried data traffic on the excess bandwidth associated with the time division multiplexed (TDM) voice networks were implemented as a lower-cost alternative to the public switched telephone network (PSTN). Thus, the concept is not new, but with the emergence of client/server networking and the Internet, this voice-based communications paradigm began to change as data traffic started to consume a far greater percentage of network bandwidth. At first, frame relay circuits for data communications proliferated, as they were much cheaper than leased lines. However, frame relay was not then well suited to voice communication, so dedicated T1 remained the channel of choice for telephony.

Carrying Voice Over Data Channels

During the mid-'90s data and voice began to merge, propelled by advances in compression technology, but this time on frame relay networks. The ubiquity of routed IP networks, and the desire to trim telephony costs are the major driving forces for the deployment of voice over IP (VoIP). One of the major advantages of VoIP technologies is that they provide a way to leverage existing network resources and to dramatically reduce, or even eliminate, telephony costs.

For instance, if a company had an IP network and branches throughout the country and overseas, there could be a voice gateway between the network routers and the PBXs at each of the sites, and voice traffic could be piggybacked on the frame relay WAN link between the cities. The immediate advantages are huge, as there would be local call rates for long distance and international calls. Even if a WAN link were not available, as long as each of the company's sites had a local point of presence (POP) for router-based Internet connectivity, the voice call could still go out over the Internet. The phone charges, as with Internet connections, would only be for the monthly service charge for the POP.

The advantages of deploying VoIP are evident. The issue of whether or not to deploy VoIP is more concerned with technical implementation and QoS than with ROI and cost-benefit analysis. The goal when implementing VoIP is to find a technology that supports telephony—including both voice and fax—over a variety of infrastructures including frame relay, ATM, IP, and the Internet.

When conducting needs assessments, make sure to include both voice and fax services, as fax is still a major business medium, regardless of e-mail. This can be tricky, as gateways need to be able quickly to identify a call as voice or fax to eliminate the need for separate ports for each medium.

Multiprotocol Support is Key

When deciding on the protocol supported, think ahead. Data communication technologies are changing rapidly. Some companies may adopt one technology versus another, based upon their own needs, a vendor's influence or even a carrier's support. The gateway solution should support all the main communication links, such as frame relay, ATM, IP, and the Internet, as they may exist between any two sites. With the flexibility to carry voice over all of these links, investments can be protected, even while all resources are leveraged. This approach also empowers selective allocation of differing communications networks for various needs, so that the network that makes the most sense for each application is chosen on a cost and quality basis.

For example, for voice calls between sites A and B that have both frame relay and IP network links, the frame relay channel could be chosen for voice because with compression, 12 voice channels can be carried in a 64-Kbps circuit. IP would be able to carry only 10 voice channels on this same circuit. However, between sites B and C that have IP and Internet channels available, IP would be favored for voice because quality tends to be higher. It would be possible to choose to send faxes over the Internet, since they have better integrity in this medium than voice does, and the bandwidth of the IP link can be optimized. To ensure optimal communications between all sites, the gateway should also feature any-to-any linking so that a voice call can come into one site via IP, be switched out over frame relay, and reach a final destination via the Internet.

Providing for Scalability and Management Tools

Configuring and maintaining a voice gateway requires a tightly integrated network management system that features a graphical user interface (GUI). This GUI should not only be easy to use, but also have the intelligence to prevent the establishment of invalid configurations. In Chapter 7 some of the major vendors and their products and services are outlined and discussed.

Make sure theVoIP implementation has large bandwidth capacity. Also, support for a direct PBX connection via a T1 or E1 should be a requirement. Gateways should be capable of supporting multiple digital interfaces and able to yield as many as 180 ports on a single gateway platform, so future growth and the ability to link multiple platforms can be ensured. Make sure that all features and functionality, including support for frame relay, IP, and Internet, voice/fax transparency, billing, and management are available to all platforms—no matter how large or small.

Vienna Systems Corp. has been a power in the VoIP market since September 1997, when they announced the first multipoint gateway to be used as a translation mechanism for sending voice traffic over corporate IP networks. The gateway could also be used to send voice over the Internet or corporate intranet. Today, Vienna Systems offers a state-of-art solution, as shown in Figure 3.1, called Vienna.way, an architecture designed to be fully distributed and flexible, making it simple to build new VoIP applications and services.

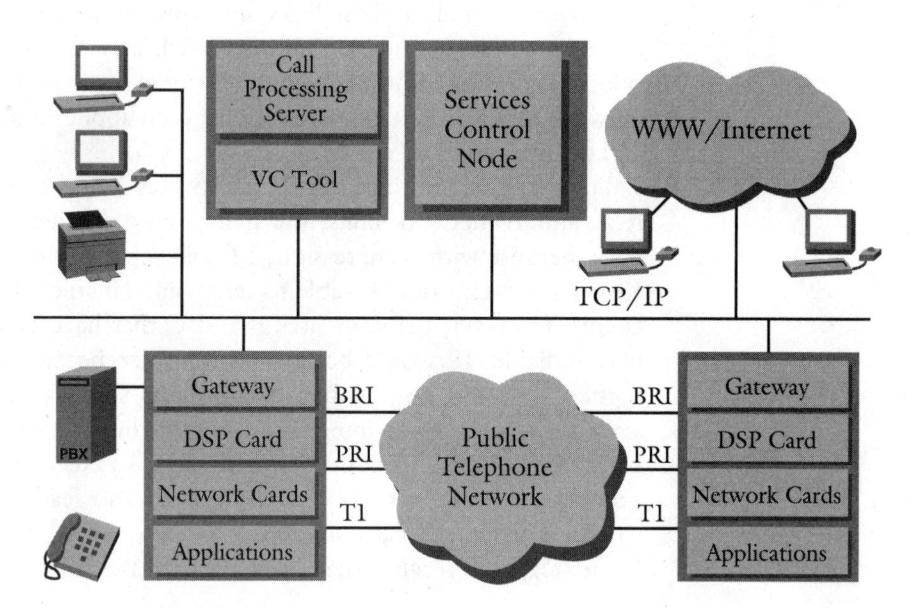

FIGURE 3.1 The Vienna.way architecture

Micom Communications Corp., a division of Nortel, leads the worldwide Frame Relay Access Device (FRAD), with 83 percent of voice ports in that category deployed, as of the fourth quarter of 1997. Micom's Marathon I-FRAD product integrates data, voice, fax and LAN traffic over a single public frame relay permanent virtual circuit (PVC) and/or private leased line from 9.6 to 1.544 Mbps or 2.048 Mbps. Intracompany phone/fax calls bypass the PSTN, eliminating long-distance toll charges and the need for multiple wide-area connections. Figure 3.2 is a screen shot of Micom's site.

The example of these two vendors gives you an idea of how demanding the VoIP industry now is. The driving force is the savings companies can realize by consolidating different types of traffic over a single IP WAN connection. At the same time, they can prepare for new Internet-based voice applications expected to become popular over the next two to three years.

Because companies' IP-based WANs often have extra bandwidth, the tendency is for intranets to start carrying the multimedia traffic. By employing products like Vienna.way, users can make voice calls to and from multimedia-capable PCs, as well as route IP-based voice traffic to and from standard telephones over the PSTN. By the same token, products such as Micom's VoIP gateway can take care of fax transmissions over the Internet, as the fax transmissions can accommodate seconds-long delays, but voice traffic, which is more delay-sensitive, should stay with the company's standard 56Kbps and 1.5Mbps leased lines.

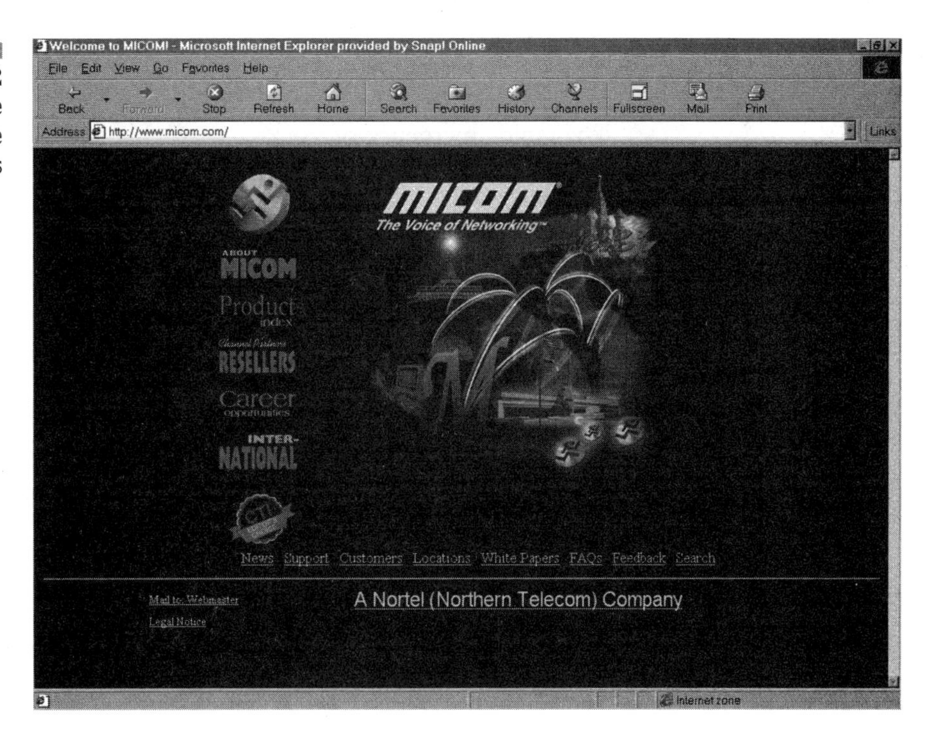

FIGURE 3.2 Micom's Web site: one of the leaders in voice over networks

Another concept gaining momentum is cable-based VoIP, which is expected to follow the lead of copper-based Internet telephony by focusing on niche services. There are still a number of issues to be resolved, including perfecting gateways between the Internet and the PSTN, developing techniques to allow Internet phone devices to ring like conventional phones, and integrating billing and management systems with those users already in the operators' system.

On the higher end, cable operators have also been consolidating ownership of systems in metropolitan areas under ever-fewer companies and deploying Synchronous Optical Network (SONET) and other technologies to tie headends together. This could form the basis of large-scale telephone networks in which cable operators provide local access through

Internet telephony and cities are connected via a combination of IP and interexchange carrier networks.

In summary, the adoption and implementation of VoIP can be justified by:

- ✔ Increasing voice/data convergence
- ✔ The fact that IP is now the "common protocol"
- ✔ Packetized compressed voice has demonstrated cost-effective solutions
- ✔ Rapid growth of intranets and extranets
- ✔ Voice over frame relay being successfully deployed in major corporate networks
- ✔ The rapid emergence of VoIP.

The H.323 Standard

The H.323 standard is an extension of H.320, which addresses videoconferencing over ISDN and other circuit-switched networks and services. H.323 is a logical and necessary extension of the H.320 standard to include corporate intranets and packet-switched networks. Because it is based on the Real-Time Protocol (RTP/RTCP) from the IETF, H.323 can also be applied to video over the Internet.

The International Multimedia Teleconferencing Consortium, Inc. (IMTC) is the organization dedicated to promoting and facilitating the ongoing development and implementation of an open, standards-based and interoperable products and services for Multimedia Teleconferencing, specifically the ITU T.120, H.320, H.323, and H.324 suites. The H.323 specification was adopted by the International Telecommunications Union (ITU) as the standard for voice and video communications over packet-switched networks such as the Internet and corporate LANs. It applies to multipoint and point-to-point sessions, in accordance with other ITU multimedia teleconferencing standards. The components of the standard are summarized in Table 3.1.

For more information about IMTC, visit their Web site at **http://www.imtc.org/u/u_search.htm**.

TABLE 3.1
Components of H.323

Recommendation	Description	Status (as of Oct.'96)
H.225	Specifies messages for call control including signaling, registration and admissions, and packetization/ synchronization of media streams	Ratified
H.245	Specifies messages for opening and closing channels for media streams, and other commands, requests and indications	Ratified
H.261	Video codec for audiovisual services at P x 64 Kbps.	Ratified
H.263	Specifies a new video codec for video over POTS.	Ratified
G.711	Audio codec, 3.1 KHz at 48, 56, and 64 Kbps (normal telephony).	Ratified
G.722	Audio codec, 7 KHz at 48, 56, and 64 Kbps.	Ratified
G.728	Audio codec, 3.1 KHz at 16 Kbps.	Ratified
G.723	Audio codec, for 5.3 and 6.3 Kbps modes	Ratified
G.729	Audio codec	Ratified

The range of networks to which H.323 can be applied, as well as the ratification of its core components is significant, and this has promoted growth in products and services based on H.323. Interoperability is also becoming critically important, as more and more H.323 developments are announced.

Lucent Technologies, through Bell Laboratories, contributed significantly to the development of the H.323 standard, and has been licensing the H.323 protocol to software developers since January 1997. Lucent's objective is to help developers accelerate the deployment of standards-based multimedia communication products on the Internet and intranets. The company is strongly committed to the widespread adoption of H.323, providing the source code and object code licenses on a variety of platforms for client and server applications.

TIP

For additional technical information and sound samples, check elemedia's Web site, at **http://www.lucent.com/elemedia/**.

Vocaltec Ltd. is another vendor supporting the H.323 standard. In Fall 1997 Vocaltec already demonstrated its Internet phone interoperability with Microsoft and Intel's telephony products. Vocaltec's acclaimed Internet Phone software now comes with H.323 embedded and it can be fully integrated with Microsoft's NetMeeting and Intel's Internet Phone software.

For more information on the VocalTec product line, visit the company's Web site at **http://www.vocaltec.com** or e-mail to **info@vocaltec.com**.

DataBeam H.323 Toolkit Series

DataBeam's H.323 toolkits provide third-party developers with the components and capabilities required for rapid building of robust standards-compliant products. The toolkits are structured to provide developers with the functionality they need for their own product plans. The DataBeam H.323 Toolkit Series includes the following toolkit offerings:

- ✔ **H.323 Core Toolkit**—A standards-compliant code base that manages call control and RTP/RTCP functionality
- ✔ **H.323 Value Pack**—A robust set of added value components for H.323 development on the Windows platform
- ✔ **H.323 Gatekeeper Toolkit**—A portable, standards-based solution for H.323 gatekeeper functionality

For more information about the H.323 toolkit, check Databeam's Web site at **http://www.databeam.com/h323/info.html**.

IP over ATM

This section was based in part on RFC 1577, authored by Mark Laubach, of Hewlett-Packard Laboratories. RFC 1577 is a standard track, which defines classical IP and ARP over ATM.

For further information about this RFC or IP over ATM, Mark Laubach can be reached at Hewlett-Packard Laboratories, 1501 Page Mill Road, Palo Alto, CA 94304. His e-mail address is: **laubach@hpl.hp.com**.

Refer to the current edition of the "Internet Official Protocol Standards" (STD 1) for the state and status of this standard. Also, as this section introduces general ATM technology and nomenclature, I recommend reviewing the ATM Forum and ITU-TS references for more detailed information about ATM implementation agreements and standards, as well as Chapter 2, which provides an overview of ATMs.

The deployment of IP over ATM into the Internet is still new; it will be some time before it is completed. However, it is already possible to see traditional IP subnet boundaries deployed over ATM. The reasons for such a strategy are outlined below:

- ✔ Systems administrators and IT professionals tend to be conservative about new technologies, following the same familiar models they've earlier deployed. Until ATM builds credibility, these professionals will always hold back, doing nothing that could compromise their corporation's ability to conduct business effectively.
- ✔ Corporate security policies often rely on the security, access, routing, and filtering capability of IP Internet gateways via routers or firewalls. However, ATMs are not allowed to "back-door" around these mechanisms, as this would jeopardize the security of the corporation. Thus, ATMs need to provide better management capability than do the existing services and practices.
- ✔ Although RFC 1577 is almost four years old, standards for global IP over ATM will take some time to complete and be deployed.

Encapsulating IP

The ATM Adaptation Layer (AAL) segments datagrams into cells, passes them to the ATM network for transmission, and reassembles the cells into datagrams at the destination. It is roughly equivalent to the data-link layer in the OSI 7-layer model. Five different AALs have been defined; the industry standard for data transmission over ATM is ATM Adaptation Layer 5 (AAL5).

AAL5 encapsulates a higher-layer datagram (such as IP) in an AAL5 datagram, as shown in Figure 3.3. AAL5 datagrams are of variable length, from 1 to 65,535 octets, plus an 8-octet *trailer*. ATM packages one IP datagram into one AAL5 datagram, segments the datagram into cells, and sets the AUU (ATM-layer-user-to-user) parameter in the last cell of the datagram to mark the end of the AAL5 datagram. Although AAL5 accepts up to 64k-sized datagrams, TCP/IP restricts this MTU to 9180 octet datagrams—IP will fragment any larger datagrams when passing them to AAL5 for encapsulation.

FIGURE 3.3
The trailer is always placed in the last 8 octets of the final cell

8-Bit UU	8-Bit CUU	16-Bit LENGTH	32-Bit FRAME CHECKSUM

The trailer is always placed in the last 8 octets of the final cell. It has 4 fields, only two of which are currently used—the length and frame checksum fields. The User-to-User Indication field (UU) and the Common Part Indicator field (CPI) were added after the initial AAL5 proposal and are currently unused.

TCP over ATM

TCP over plain ATM does not have a record of good performance. It can be significantly worse than standard "packet TCP". This is because of segmentation of TCP/IP packets at the ATM layer, since TCP packets are segmented into many 53-byte cells by the AAL5 layer. Any loss of a single cell causes the effective loss of the whole TCP packet. Also, in a manner unlike that of the traditional packet-switched networks, when the TCP packet is corrupted by the loss of a single cell, the rest of the cells are still forwarded to the destination, clogging the congested link with useless data. This situation can worsen if any of the factors listed below contributes to the increase in the number of cells dropped at the switch:

- ✔ Small buffers
- ✔ Large TCP packets
- ✔ Inefficient TCP window size
- ✔ Increased number of active connections.

Buffer sizes usually range from 256 to 8000 cells (per port) and a buffer size of 1000 to 2000 cells for a small switch (16–32 ports) is common. The use of larger packet or window sizes increases the number of wasted cells that the congested link transmits when the switch drops a single cell from one packet.

Thanks to partial-packet discard (PPD), also known as selective cell discarding (SCD), if there is congestion caused by the drop of a cell, the PPD will attempt a traffic control mechanism by dropping all subsequent cells of the packet in question. Once the switch drops a cell from a VC (Virtual

Connection), the switch continues dropping cells from the same VC until the switch sees the AUU parameter set in the ATM cell header, indicating the end of the AAL packet.

This type of congestion control can be established on a per-VC basis (for AAL5). PPD requires the switch to keep additional per-VC information in order to recognize which VCs are using AAL5 and want to use PPD. It must also maintain a record of which VCs are currently having cells dropped.

PPD offers a limited improvement, however, because the switch begins to drop cells only when the buffer overflows. The first cell dropped might belong to a packet in which the majority of cells have already been forwarded. Also, when the switch first drops a cell, the switch does not look in the buffer for earlier cells that belong to the same packet. Thus cells from the corrupt packet may be forwarded from the switch even though PPD is in effect.

TCP's default-clock granularities represent another limitation of TCP over ATM, as they are inappropriate for a high-speed ATM network. When a TCP packet is dropped due to congestion, the retransmit timer gets set to a relatively large value, compared to the actual RTT. A TCP clock granularity of .1 milliseconds works well for TCP over ATM. However, newer versions of TCP like TCP Reno and Vegas have improved congestion control algorithms that reduce reliance on the retransmit timer.

Voice over ATM

IT professionals, more specifically systems managers directly involved with ATM carrying voice and data over the same network, are disappointed. After all, the general expectation in the ATM industry is that the technology exists, is proven and works. In reality, voice over ATM has turned out to be a more expensive and inefficient way to carry voice than TDM over leased lines.

According to the ATM Forum, voice should be transmitted as CBR (constant bit-rate) traffic. CBR is a method that forces customers to reserve bandwidth for voice even when they are not actually sending voice. If voice is sent as VBR (variable bit rate), even though this may sound an obvious alternative (as it permits allocation of voice bandwidth on an as-needed basis, reducing costs with voice calls), trouble may ensue.

VBR for voice is not yet standard, even though the ATM Forum is currently evaluating ways to write it into a revised ATM specification. Be careful with the "pre-standard" VBR equipment which may be available. The problem the industry faces with voice over ATM is more a specification issue than anything else. Furthermore, any pre-standard specification is proprietary, and that preduces a voice-over-ATM double bind, as it forces investment in standards-based but inefficient CBR products, or the sacrifice of interoperability for savings by going with VBR offerings.

IP over SONET

In carrier and ISP industry, many carriers are looking for ways to leverage their existing investments in SONET setups by offering IP services directly over them and avoiding the bandwidth overhead normally associated with ATM.

The market for IP over SONET is growing, as is the market for IP over ATM, especially overseas. This means that IP traffic is poised to become the ubiquitous dialtone for a host of services that extend beyond the transport of HyperText Transport Markup Language (HTML) traffic and flat image files. I believe that the volume of IP traffic will easily take over voice on carrier networks within the next couple of years, and the nature of that traffic is currently shifting to multimedia applications, such as electronic commerce, and those delivered by publishing and subscription technologies.

SONET has turned out to be one of the greatest surprises in communications technology of the '80s. Conceived in 1984 as an optical network standard that would allow disparate network elements to interface, SONET addressed a specific concern of local carriers: a fundamental, midspan cable need. SONET four years ago was nowhere near the big hit it is becoming today with its huge bandwidth capacity. Actually, SONET broadens the bandwidth concept so much that the distinctions between transmission, switching, and CPE become blurred. Further, SONET has the capacity to combine separate voice, data, and video networks into one broadband, multimedia network, by encapsulating in fixed-length, variable-position cells transported by super high-speed synchronized frames over the SONET network.

Do not be surprised if the gigabit-transmission rates of SONET begin to rival the internal bus speed of many mainframe computers, which will turn SONET networks into a wide-area bus for the computer, assuming the role of a full-function server in the ideal distributed-computing environment.

Many industry analysts believe ATM-based switching, combined with SONET-based transport, is the network solution that will eventually dominate. Local and interexchange carriers are abandoning T-3 in favor of SONET as they expand. In times of such a data communication technology blur, be sure to stick with industry standards when designing broadband networks.

SONET Benefits

SONET offers tremendous benefits to both telephone carriers and end-users. It was conceived and created as a network, not simply a transmission path or piece of network gear. Consequently, SONET is an end-to-end service designed to satisfy the following communication needs:

- ✔ Automated maintenance and testing
- ✔ Bandwidth administration
- ✔ Real-time rerouting
- ✔ SONET digital cross-connecting
- ✔ Standard optical interface
- ✔ Synchronous multiplexing
- ✔ Worldwide connectivity.

SONET synchronous transmission offers the capability of direct access to individual DS-0 and DS-1 channels. This provides a simple and effective means to achieve automated control over individual voice channels. What is more, this control can take place in "real time", providing true "bandwidth-on-demand" capability.

SONET enables carriers to tailor the width of the information highway. This can be done from a remote location, allowing carriers to respond expediently to the specific needs of their customers. This remote provisioning can result in the ability of carriers to provide new services, such as network reconfiguration.

SONET also has the ability to support real-time rerouting, allowing customers to bypass congested nodes or points of failure by reconfiguring the routes of affected circuits. The reroutes would be predetermined and stored in Automatic-Call-Routing (ACR) programs to provide immediate network recovery. SONET'S automated maintenance and testing capability through embedded control channels enables carriers to track the end-to-end performance of every transmission. This capability allows carriers to guarantee transmission performance, and users can readily verify compliance through on-premise management terminals.

Synchronous Multiplexing

SONET supports transmission rates from 51.84 Mb/s through 2.488 Gb/s. Synchronous multiplexing allows the high-speed transmission element within the multiplexer to observe and extract the lower-speed digital signals. This mode of operation allows add/drop time-slot interchange multiplexing without bringing all the signals down to the DS-1 level, which enables the elimination of mid-level network elements, such as back-to-back M13 multiplexers.

Today's T-3 transmission equipment lacks a universal protocol standard and therefore DS-3 signals are proprietary, varying among vendors. Accordingly, these DS-3 schemes are not compatible wiath each other. A user is required to employ a particular manufacturer's equipment at both ends of a link. SONET alleviates all this and allows mid-span meet, which is the ability to interconnect equipment from different vendors.

SONET promises true interconnectivity between fiber transmission equipment vendors. Users as well as carriers will then have more choice in product selection, making it far easier to implement solutions. Carriers and users will be able to purchase equipment based on price and performance, and mix and match hardware from multiple vendors as needs warrant.

SONET heralds a fundamental change in the network from basic physical connectivity to integrated end-to-end administration and maintenance. It will be a catalyst for the modernization of the network, both private and public. SONET, along with ATM, will transform the network from a circuit-switched, to a high-speed cell-switched, dynamically robust infrastructure.

How quickly SONET deployment evolves will depend more on politics than on technology. The FCC's aggressive tactics to promote competition and deregulation may be the key impetus. Increasing "local loop" competition from Competitive Access Providers (CAPs) and cable television providers is driving the deployment of SONET equipment in LEC operating territories. Already faced with competition, the IXCs are rapidly deploying SONET equipment as well.

Voice over SONET

The Cirrus product family is the industry's first series of products to use a thin layer of ATM over SONET/SDH fiber to merge transmission, access and termination into one integrated access system. The thin layer of ATM multiplexing ensures that the network efficiently delivers traditional voice services and can support advanced services such as VoIP, symmetric and asymmetric Digital Subscriber Line (DSL), and packet over SONET (POS).

Ten years ago, ISDN was going to spark a revolution in which users could speak and transmit data simultaneously over the telephone. SONET and ATM were already in the pipeline, promising highly integrated voice, video and data transmission rates. Today, even though ATM and SONET are pitted against each other in the race to deliver ever-higher network bandwidth, ATM and SONET technology are more frequently teamed. Aside from data rates starting at OC-1, or 51 Mb/sec, and scaling to OC-

192, or 9.9 Gb/sec, SONET brings its self-healing ring architecture, which provides unparalleled survivability in the event of a fiber cut or node failure. If such a break occurs, SONET is able automatically to reroute data in the other direction around the ring.

SONET and Other Data Streams and Protocols

The design of the frame and signaling for SONET makes it compatible with the traditional, existing networks. Most prominent of these is the telephone network. The frame format described above for STS-1 was chosen so that the 125 microseconds to transmit matches the standard telephony 64 kilobytes per second circuit. A single STS-1 payload is capable of carrying 672 voice channels. Table 3.2 shows a comparison of SONET with the existing digital signals and voice channels.

TABLE 3.2 SONET verus existing digital signals and voice channels

SONET	Data Rate (Mbps)	DS-0s (64 kbps)	DS-1s (1.54 Mbps)	DS-3s (2.048 Mbps)	Voice Channels
STS-1	51.84	672	28	1	672
STS-3	155.52	2016	84	3	2016
STS-9	466.56	6048	252	9	6048
STS-12	622.08	8064	336	12	8064
STS-18	933.12	12,096	504	18	12,096
STS-24	1244.16	16,128	672	24	16,128
STS-36	1866.24	24,192	1008	36	24,192
STS-48	2488.32	32,256	1344	48	32,256

Ideally, there should be ATM over a SONET infrastructure, as shown in Figure 3.4. There can be DS3-based (45 megabits/sec) ATM networks, but there is a lot of overhead with ATM, and the efficiency of those networks is not as great as the higher-speed SONET's. ATM and SONET's proven combination of speed, reliability and flexibility have also enabled early adopters within the government to tackle a host of applications stymied by the bandwidth limitations of previous network setups. Those applications typically rely on multimedia capabilities ranging from compute-intensive imaging applications to distance learning, videoconferencing and even ultra-high-speed Internet access.

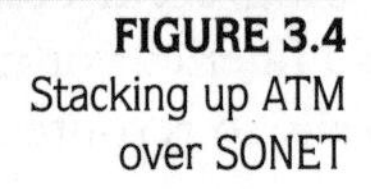

FIGURE 3.4 Stacking up ATM over SONET

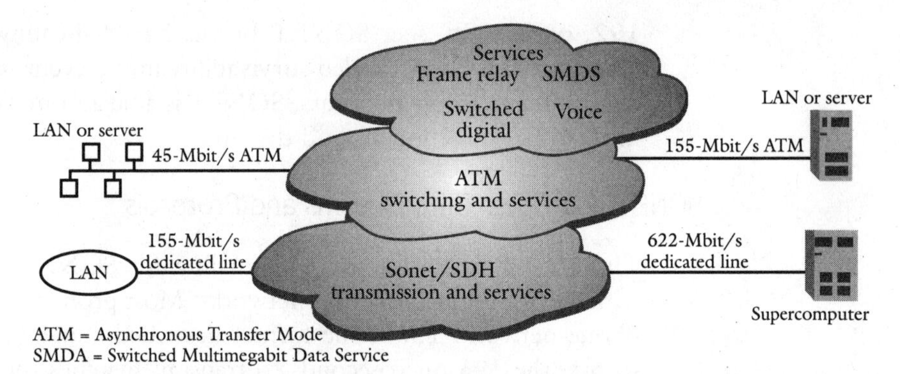

However there are problems in trying to marry voice over ATM/SONET. On the ATM front, installation can be complicated, and there is not much available for managing the network. ATM's technology is still new, lacking much experience, as the pioneers are still building their learning curve and the "do's" and "don'ts" list. SONET has its hurdles too; scaling it up is a very expensive process.

However, there are many vendors climbing the SONET technology, making tremendous progress, and making sure that it soon will be ready for prime time. Fore is building SONET-like restoration into its ATM switches via the Private Network-to-Network Interface (PNNI), a standard for dynamically rerouting traffic, and via SONET automatic port switching.

IP over Frame Relay

Recent technologies have brought packetized voice, or the transmission of voice and fax over IP and frame relay networks, into the mainstream. An understanding of the key technologies is essential for effective, high-quality implementation of enterprise solutions.

Early packet networks were based on X.25 and other proprietary statistical multiplexing protocols running over modem or low-speed digital circuits, which had to cope with substantial overhead and delay.

It was only in the early 1990s that frame relay became more common and started to threaten the position of X.25. Frame relay takes advantage of the higher reliability of modern digital networks to carry packet data with reduced error checking and retransmission, over higher bandwidths.

The growth in frame relay since that time has exceeded all expectations, with thousands of companies turning to this efficient, cost-effective service to fulfill data communication requirements. Frame relay now plays a significant part in many networks' infrastructures, and companies are beginning to migrate voice and facsimile applications over their frame relay networks. By migrating to frame relay, companies have further consolidated their networks, gaining the cost advantage of fewer WAN links without compromising their ability to transport voice, fax and data traffic effectively.

The IP is not mutually exclusive to frame relay. In fact IP runs over frame relay, PSTN and many other types of networks. Like frame relay, however, IP networks are packet-based. Therefore the advantages and limitations inherent in IP voice are part of the overall packetized voice discussion.

IP introduces even more performance challenges than does frame relay for voice applications. However, IP has become ubiquitous as a result of the Internet/intranet explosion. And, beyond its ubiquity IP offers a new dimension of value in terms of voice-data application integration.

Voice over Frame Relay

Every so often, the advantages of implementing voice over a frame-relay network, especially for corporations already having voice available on virtual private networks (VPNs), are not easily identified, and IT professionals base their decisions on economics.

This section was based on Nuera Communications' white paper by Steven A. Taylor. Nuera is one of the major players in VoIP. For additional or more detailed information, or for products that support it, check Nuera's Web site at **http://www.nuera.com**.

Voice over frame relay can save money, since it is as inexpensive as half a cent per minute. Actually, even if voice over the public networks cost as little as five cents, voice over frame relay can often pay for itself in less than a year. Nonetheless there are many alternatives to transport voice, including leased lines, ATM and the Internet. Even though these are all viable technologies for packetized voice transport, there are a number of factors that make frame relay an excellent choice today.

Specialized "Voice FRADs" are readily available both for voice transport and for transporting other data along with the voice. These other data could include both LAN internetworking via routers and more-traditional FRAD functions for serial protocols like SNA and X.25. The main challenge in adopting voice over frame relay is technical; it will often be necessary to surmount various technical hurdles in order to tune implementation and ensure good voice quality.

Voice Over Frame Relay and Cost Benefits

Voice over frame relay can be the right solution for controlled environments where all of the voice can be considered to be on the protected network (LAN/WAN). It is most easily justified for intracorporation communications between sites on the corporate enterprise network, using the same facilities that are already in place for the corporate data infrastructure. Similarly, a carrier who wishes to provide voice services for users connecting over its internal network can reasonable justify it.

As security technologies, including encryption schemes, become available, packet voice (over the Internet, frame relay, or ATM), may also be a reasonable alternative for applications outside the boundaries of an intranet, whether both parties are using the packet service or not. Until then, the primary application for voice over frame relay will be the same as that for which we have been using internal voice communication for over ten years, with multiplexers providing virtual tie lines between PBXs attached to the network. The voice capabilities are then used both for communications between LAN/WAN connected sites and for remote dialing, as shown in Figure 3.5.

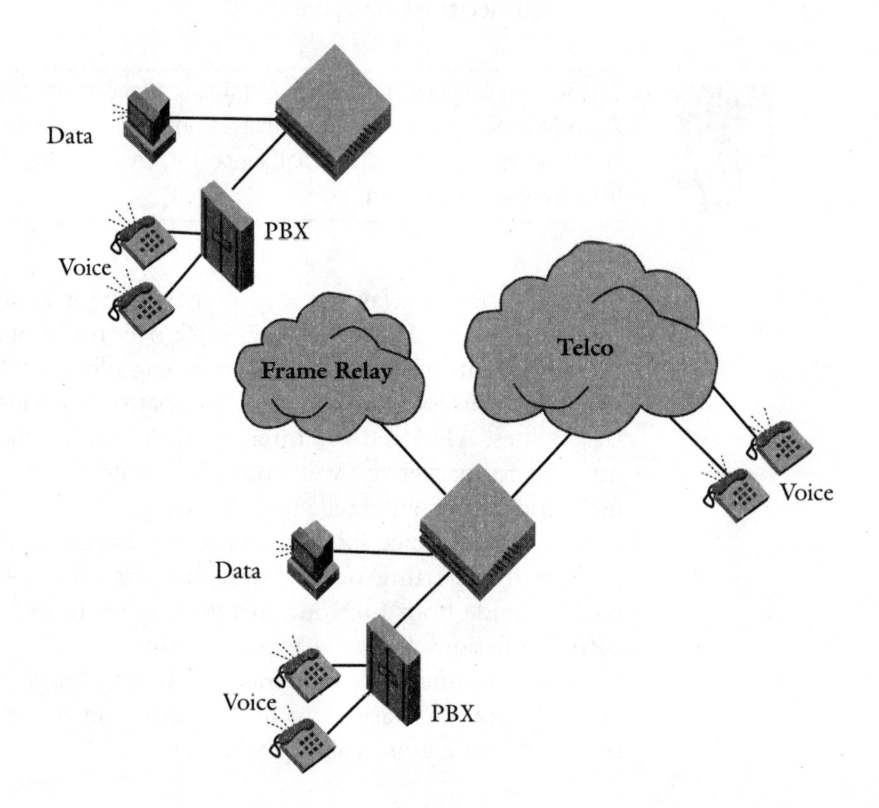

FIGURE 3.5 Intra-company voice networks

As Figure 3.5 shows, the network is used to connect to a site in a remote location. Then, at the remote location, a secondary dial tone is provided by the remote PBX to enable dialing a local call, as opposed to using the public network across the country or even around the world. This strategy alone can save a corporation much money. A call from company headquarters in California to London would be a local call, and the frame-relay voice could even take care of doing the local dialing at the remote location so that the second dial tone would never be heard.

Using voice over frame relay involves some fundamental assumptions. First, frame relay voice uses an advanced compression algorithm, such as Code Excited Linear Prediction (CELP). Advanced algorithms are made possible by extremely high processing power toll quality, with voice available at 8 to 16 kbps, as opposed to the 64 kbps required by Pulse Code Modulation (PCM).

Furthermore, it is assumed that Voice Activity Detection (VAD) is used. VAD takes advantage of the fact that normal conversations are half-duplex. That is, the transmission takes place in only one direction at a time, with silence in the other direction. As a rule of thumb, it is not a bad assumption that 50 percent of the time there is silence on the line. In reality, even more is usually silence, especially if we consider pauses between words and sentences. However, the packetization process adds a little overhead, so the silence greater than 50% is roughly canceled out by the additional overhead.

The combination of these factors represents a compression ratio of at least eight-to-one compared with traditional voice. This is calculated by assuming the use of 16 kbps for the voice itself, providing for four-to-one compression. This is doubled, though, by using VAD for an additional two-to-one compression—hence the ballpark ratio of at least eight-to-one.

The eight-to-one ratio, a conservative one, is used here simply to be conservative when evaluating financial considerations.

Comparing Dial Voice Costs with Frame Relay

Voice over frame relay is fairly well accepted as economical when used in international and non-US applications. With international call prices sometimes approaching $1 per minute, the payback period for voice over frame, even with the higher international rates for frame relay, is dramatic.

One way of looking at the economic justification for voice over frame relay involves the dollars per month based on the cost of dial voice. For instance, assume the use of only one hour each business day for each of eight lines at $1 per minute; the monthly cost is almost $10,000. When dealing with this type of cost up front, voice over frame relay is cost-justified very quickly.

The economic justification is much more of a challenge in domestic networks. For instance, assume the use of three hours on each of eight lines at five cents per minute for 20 days a month; the cost is $1440 per month for all eight lines—significantly less than international phone bills.

Voice over Private Frame Relay

Even though voice over public frame-relay services makes a lot of sense for international applications, frame-relay services are not universally available. Thus for international locations where frame relay is not available, the best solution is to use private frame-relay services running over leased lines. Furthermore, installations using public and private frame relay should be totally interoperable with appropriate voice over frame relay equipment.

In order to use voice over private frame-relay networks within the US, bandwidth must be available, so all of the bandwidth considerations mentioned above still apply. However, it is more likely that the private frame-relay network rather than the public network will have "free" bandwidth available, as public frame-relay service tends to be quite granular already. Since services are purchased in DS0 (64 kbps) increments, and even finer granularity below 64 kbps, it is possible to fine-tune the bandwidth utilization to a more precise degree than would be necessary to determine bandwidth needs. Consequently, the "excess bandwidth" that tended hang around in T1/E1 multiplexer networks should not exist, although these limits do not obtain for private frame relay networks.

One could argue that private frame-relay networks can be fine-tuned to the same extent as public services since fractional T1 is likewise available on a "by-the-DS0" basis. The interesting part comes with tariffs. Fractional T1 prices tend to increase fairly linearly through about half a T1. Once the 768 kbps threshold is reached, the price increase drops drastically. In fact, the price for a full T1, without discounts, is generally only about 20 percent greater than the price for half the bandwidth. Furthermore, since carriers tend to discount full T1 circuits much more heavily than fractional T1 circuits, we often find that a full T1 is actually less expensive than half a T1.

When this is the case, and if the private frame-relay network has internodal connectivity needs in the range of 512 kbps to 768 kbps, it could very well turn out that full T1s are as inexpensive as any other option for connecting the nodes. The result is that there is indeed "free" bandwidth available for the frame relay voice traffic. Regardless of the price per minute offered, it is difficult to beat "free" in any economic analysis.

The bottom line is that the use of frame-relay voice can easily be justified on an economic basis. Rather than hindering communications in any way, the availability of low costs due to the advanced compression techniques can actually enhance business communications. This is true for U.S. domestic and international applications for both public and private frame relay.

Is Voice Over Frame Relay a Viable Option?

Packetized voice can be transported by a wide variety of options from traditional leased lines to ATM. The main strengths of frame relay lie in its availability, its pricing, the target speed range, and the reliability of the technology. But is voice over frame relay a viable option?

Frame Relay vs. Leased Lines

There are circumstances where the use of leased lines is a solution much superior to frame relay. For instance, leased lines are a better solution when requisite data transport speeds fall in the range between 512 kbps and a full T1. In these cases, the equipment used to support voice over frame relay still works. It just becomes a special case of frame relay being run in a point-to-point environment (without the frame-relay network in the middle). However, since frame relay is pretty much available throughout the U.S., its pricing is much more attractive than that of leased lines. This becomes especially important when the network topology involves either meshed or star connectivity among a number of sites.

Frame Relay vs. ATM

ATM is another option for transporting packetized, compressed voice. As a sister technology to frame relay—with the only meaningful difference being that fixed-length rather than variable-length packets are used—most of the reasons why frame relay is good for packet voice apply equally well to ATM.

The main disadvantage of using ATM is that it is not as readily available as frame relay. While ATM services are rolling out and availability will continue to increase over the next several years, there are not yet enough ATM services available at T1/E1 speeds. ATM also has some bandwidth inefficiencies, as inherently ATMs have an overhead of at least 10 percent, based on the 5-octet header for every 48 octets of payload. Thus, for highly compressed voice, this disadvantage can be noticeable, especially since packet voice packets tend to be quite short, on the order of 10 to 15 bytes per packet (due to the need to provide samples on a regular basis in order to control delay). If we assume one voice packet per cell, well over half of the cell has to be stuffed with electronic plastic peanuts to fill out the 48 octets of payload. The result: over 200 percent overhead.

ATMs were developed for speeds of T3/E3 and above; they never worked well for T1/E1. The cell overhead is almost insignificant when considering speeds of T3 to OC3, 45 Mbps to 155 Mbps. But the minimum of eight-to-one compression provided by advanced voice algorithms makes the number of channels supported at these speeds much higher than any but the largest users need. For instance, since a T3 normally carries 672 DS0s, it would easily carry over 5,000 calls at eight-to-one compression. By contrast a T1 supports almost 200 simultaneous calls, in most cases still a reasonable number for most companies.

The net result is that for cell-based ATM, the technology is readily available for carrying packet voice, but it is overkill by the time ATM's target speed range is reached.

Frame Relay vs. the Internet

Needless to say, TCP/IP via the Internet is widely adopted. In fact, the idea of voice over the Internet is drawing considerable attention, as this book indicates. However, there are major drawbacks for Internet voice, essentially the same as those of X.25: the robustness of the protocol with its requisite delay, and delay variability. While these are problems that can be addressed with frame relay, they are more difficult to address with TCP/IP. For most applications, using the same transport layer that is already carrying TCP/IP traffic to carry voice will result in a more robust implementation since issues like queuing traffic can be addressed directly.

Equipment for Voice Over Frame Relay

The utilization of voice over a frame-relay infrastructure requires certain equipment at a site. The most common is usually called Voice Frame Relay Access Device (VFRAD), as shown in Figure 3.6.

A VFRAD is very similar to a normal FRAD in functionality, but it is specialized in that it handles the voice packetization and compression process. Of course, FRADs also have other features, such as support for SNA and possibly some other serial protocols, as well as router support, which are provided either integrally or via external routers.

The VFRAD will typically be a point-to-point type of device with a single frame-relay interface, but it can usually support multiple logical connections to a number of sites on the frame relay network by routing voice calls over several different virtual circuits. The VFRAD may also support at least one additional frame-relay interface, which is very important when utilizing existing routers in conjunction with the VFRAD, allowing the

VFRAD to perform frame switching. Furthermore, data connections for existing routers and frame-relay-accepting direct attachments can also support devices. Whether these capabilities are external or integral, the VFRAD must use priority and "fairness" algorithms to ensure that appropriate access to the network is available for all traffic types.

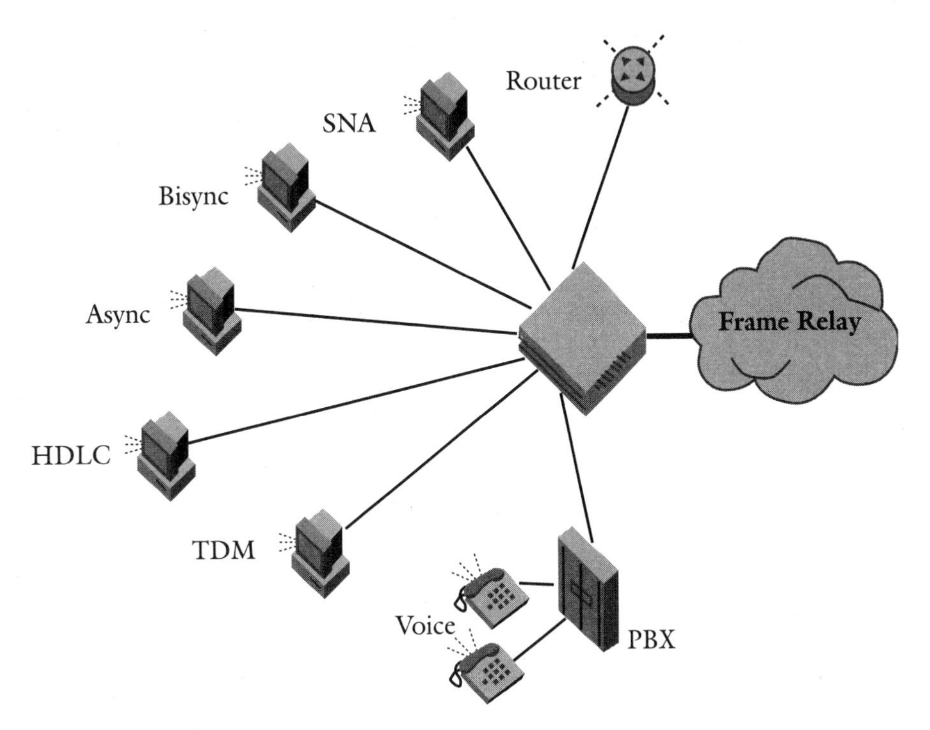

FIGURE 3.6 Typical frame relay voice implementations

The VFRAD typically supports a variety of voice interfaces. The most common is an analog interface, in which there is a direct conversion within the VFRAD from analog voice to compressed digital voice, with the compressed voice transported via frame relay. Voice trunks from the local PBX (or telephone) are attached directly to the VFRAD. The VFRAD supports the most common analog voice functions, including support for a wide variety of 2-wire and 4-wire interfaces. A VFRAD also supports digital T1/E1 interfaces for headquarters and regional sites requiring a higher density of voice connections.

Technical Challenges for Transporting Voice Over Frame Relay

Transporting voice over frame relay has its challenges. Historically, frame relay has been developed and sold primarily as a data transport technology and solution for service, not voice. This does not mean that frame relay has a fundamental technical limitation, and all of the technical challenges of using frame relay to transport voice can be met.

There may be carriers who will discourage the use of voice over frame relay. Conversely, there are carriers that actually welcome voice traffic. The reason for discouragement is that some carriers have a larger installed base of voice to protect from possible erosion by frame relay. In some cases these carriers may also be adhering to corporate policy rather than legitimate technical concerns regarding transmission capabilities.

From the technical side, not every carrier is prepared to support voice over frame relay, as the chosen switching system may not provide the capabilities and types of technical support needed by frame relay voice. The frame-relay specifications are User-to-Network Interface (UNI) only. The actual network transport is not specified. Thus, various switching architectures and network implementations will provide different levels of support for voice over frame relay.

An alternative to consider is to transport voice over a private frame-relay network, which is intrinsically easier, since the network infrastructure is under direct control. In this case, the various characteristics of the networking architecture must be sufficient to provide the desired level of voice support. But the challenges will still be there, and regardless of using main carriers or private frame relays, they must be resolved. Let us take a look at some of the most important ones.

Controlling Delay

The major challenge in transporting voice over frame relay is the use of frames itself, which generates delays. These delays, also know as absolute delays, must be controlled in order to avoid interfering with the normal human communications process. If there is more that about 500 msec total (round trip) delay, carrying on a "normal" conversation is difficult. The first level of delay is based on the "freeze-out" phenomenon. During the time that a given frame or cell is occupying the transmission facility, other traffic may not be transported.

This concern has some basis when the data frames are quite long and the speeds are low. The actual freeze-out time per node or link is visualized as an equation.

```
time = maximum frame length / transmission link speed
```

The maximum frame size can also have a major impact. If the frame size is limited to roughly 500 octets, the maximum freeze-out time is only 12.5% of the time with a 4,000 octet maximum. At the same time, frame relay still represents almost a ten-fold increase in efficiency vs. ATM. Further, the "frame vs. cell" arguments really apply only at the UNI. The intranetwork architecture is not subject to the standards, so the transport may indeed be over an ATM infrastructure. This challenge is further met by the capabilities of some VFRADs of performing sub-segmentation of data frames. If data come into the VFRAD via a frame-relay interface from a router, the maximum frame size may be too long for guaranteed excellent performance.

Consequently, the VFRAD will segment the frame into multiple frames with a shorter maximum frame length. The receiving VFRAD will then reassemble the frames. This process is analogous to the need for frame segmentation when transporting very long frames from protocols like IP over Ethernet, which has a maximum frame length of about 1500 bytes.

Worse than an absolute delay is delay variability. When a frame relay replays a voice at the receiving end, the receiver should hear a continuous and smooth talk spurt. However, packet systems are inherently prone to variable delay; the voice packets do not always get through quickly enough, and many times may be delayed a bit more.

To resolve this problem, one alternative is to buffer the end device, smoothing out minor variations. A nominal delay is intentionally introduced so that there is a high probability that all of the packets from the talk spurt will arrive in time to be played out smoothly.

Another alternative is to limit the maximum size of the frames; using cells within the transport network helps. Freeze-out is minimized and another delay factor, the fill time (the time that it actually takes to fill a frame or cell with voice), is significantly limited by use of short frames. Advanced algorithms like CELP pack a lot of information into a few bytes. Consequently, in order not to induce unacceptable delay, the frame sizes for the voice frames in addition to the data frames must be kept relatively short. Typical compressed voice frames are usually on the order of 10 to 15 octets per frame. Also, for data applications, a FIFO (First In, First Out) scheme is often considered optimal. Voice is much less tolerant of delay than data. Consequently, some form of priority queuing, both in access devices and within the switching network, will greatly enhance the probability of successfully implementing voice over frame relay.

A further delay and queuing issue to be considered is the impact of "lost" frames. For data, "late" is indeed better than "never" as far as delivery is concerned. The same is not necessarily true for voice. Some voice algorithms first send the "basic" voice information. Then, if additional bandwidth is available, enhancement packets are added. These "enhancement packets" can be sent as "discard eligible," and their loss degrades the voice quality but does not make the call unintelligible. Whether the voice packets are being transported over frame relay, ATM, or dedicated transmission services, there is a possibility that packets will be "lost" due to link errors and other factors. Consequently, processes have been built into the advanced algorithms to compensate for a certain degree of packet loss. Parenthetically, the short frame sizes that are an advantage for controlling delay also help minimize the impact of lost frames since there is less information per frame.

The bottom line for queuing algorithms is that unlike with data frames that should be delivered whenever possible, there comes a point when it is better to lose a voice frame than to deliver it too late. Consequently, it is both possible and necessary to consider the use of voice when looking at various intranetwork-buffering schemes.

Fax and modem traffic present a much larger technical challenge. They are treated together here since faxes tend to use off-the-shelf modem technology for the actual fax transmission. The good news about fax machines and modems is that they are incredibly inexpensive and virtually everybody has them. The bad news is exactly the same. Within the corporate environment, the purchase and use of these devices is often not under the control of the corporate information-processing department, so even knowing when and where these devices might exist on the intracorporate network may be impossible.

Fax and modem traffic is normally transmitted over "telephone" lines. The modulation schemes used for these devices assume the nominal bandwidth and other characteristics of "real" uncompressed phone lines using traditional 64 kbps PCM voice modulation schemes. Highly compressed voice, while sounding very good for speech, uses different algorithms that do not necessarily support the data modulation algorithms. For instance, most advanced modem modulation schemes include "phase shifts" supported by traditional high bandwidth (and analog) voice techniques. However, the human ear cannot detect phase shifts, so preserving phase-shift information does not add to the quality of highly compressed voice. Consequently, most highly compressed voice algorithms do not support modem traffic at speeds over 4.8 kbps. In order to support higher-speed modem and fax traffic, which is quickly becoming the norm, one of several approaches must be taken.

The first and simplest approach is to declare that modems and fax machines should not be used over the private, intracorporate voice network. This requirement is likely to be politically unappealing. After all, shouldn't an "advanced" voice network be able to support all of the "legacy" applications, whether they're literally "voice" or not? Further, this approach would require that the "voice" network still have some lines for on-net traffic for intracorporate fax and modem traffic. This bifurcated corporate voice network then tends to start losing many of the advantages we hope to gain by using voice over frame relay.

Beyond this segregated approach, there are two additional approaches that offer technical solutions to this historically difficult problem. The technically simpler of these two approaches is to turn off the compression whenever fax or modem traffic is detected. Basically, the equipment listens for fax and modem tones, and whenever they are detected a full 64 kbps is dedicated to the voice call rather than the compressed bandwidth. The advantage of this approach is that it works and is simple. The disadvantage is that it is a "brute force" approach that uses more bandwidth rather than less.

Another approach, as shown in Figure 3.7, is to terminate the analog portion of the fax or modem traffic at the VFRAD, transport the information as data at the appropriate rate, then remodulate as an analog call at the remote VFRAD. This way, the call takes only the amount of data really needed to transport the digital information. Note that there is very little traffic sent by the receiving fax.

FIGURE 3.7
One fax and modem solution: remodulation

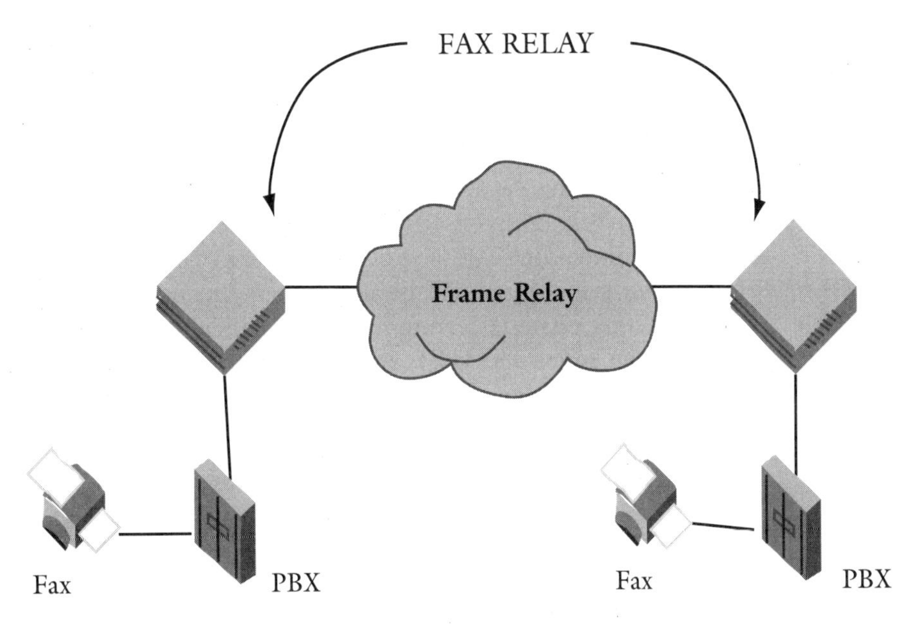

Another challenge for highly compressed voice is avoiding multiple tandems. As depicted in Figure 3.8, it is fairly common for calls to be passed through multiple switches. However, highly compressed voice does not fare particularly well through multiple compression/decompression cycles, so whenever possible the calls should be limited to a single compression.

FIGURE 3.8
Multiple tandems

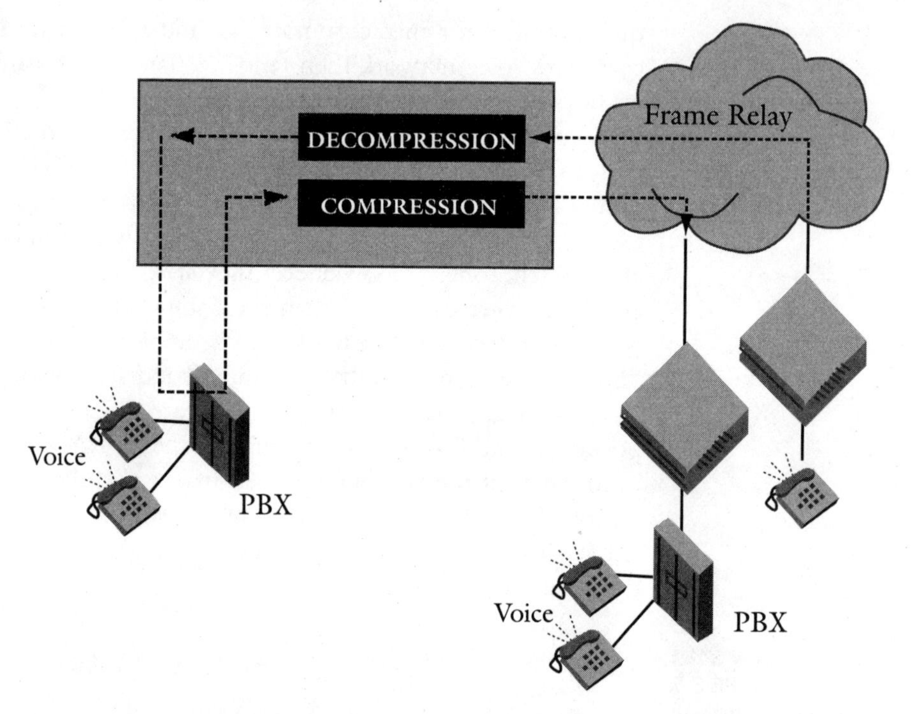

Figure 3.9 shows one solution to this problem, performing frame switching within the VFRAD. In this situation, the VFRAD has sufficient frame-switching capacity to recognize that some calls may not physically terminate at a given site, so the frames are "bypassed" to their ultimate destination. The obvious advantage of this solution is that the voice stays in the compressed digital format throughout the network with a single compression/decompression. Further, the requisite number of PVCs for the entire network is smaller since direct connectivity among all sites is not needed. Nevertheless, there is the drawback that the traffic must traverse the frame-relay interface twice (into the VFRAD and back out) at the intermediate site, adding delay.

An alternative solution, shown in Figure 3.10, avoids multiple passes over frame-relay interfaces by using full network connectivity for PVCs between each of the sites. Each call is "routed" to the appropriate destination by interpreting the DTMF "dialing" tones as the call is established, and, based on the "phone number," the call is placed on a particular PVC. Once again, the advantage is that multiple compression/decompression cycles are avoided. Staying within the network minimizes delay. Nevertheless, this configuration usually requires more PVCs than the method in Figure 3.14. Also, this requires that the VFRAD be more sophisticated to handle all of the dialing interpretation functions.

FIGURE 3.9
Switching in FRADs

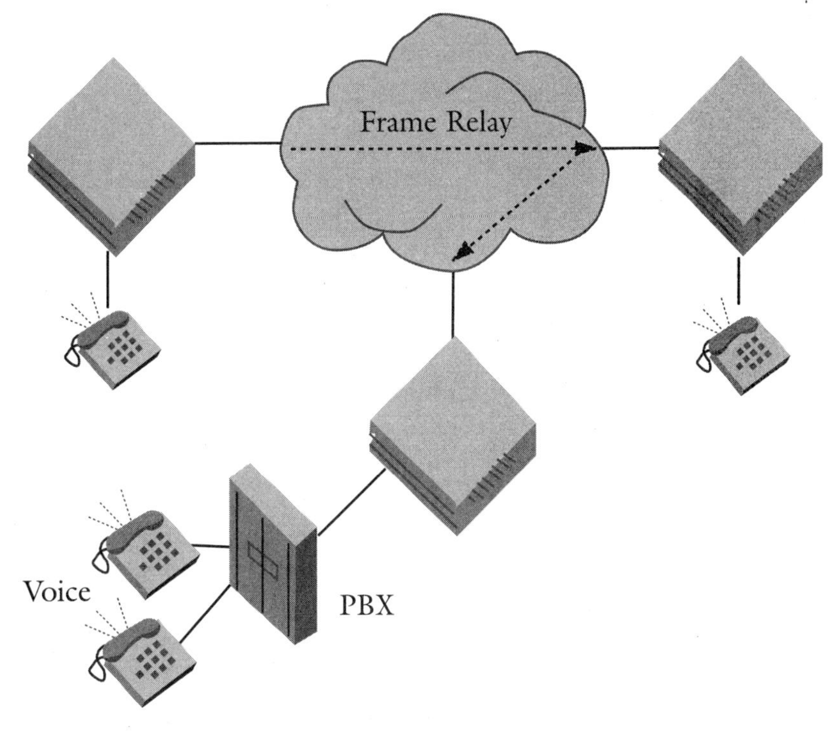

FIGURE 3.10
Call processing

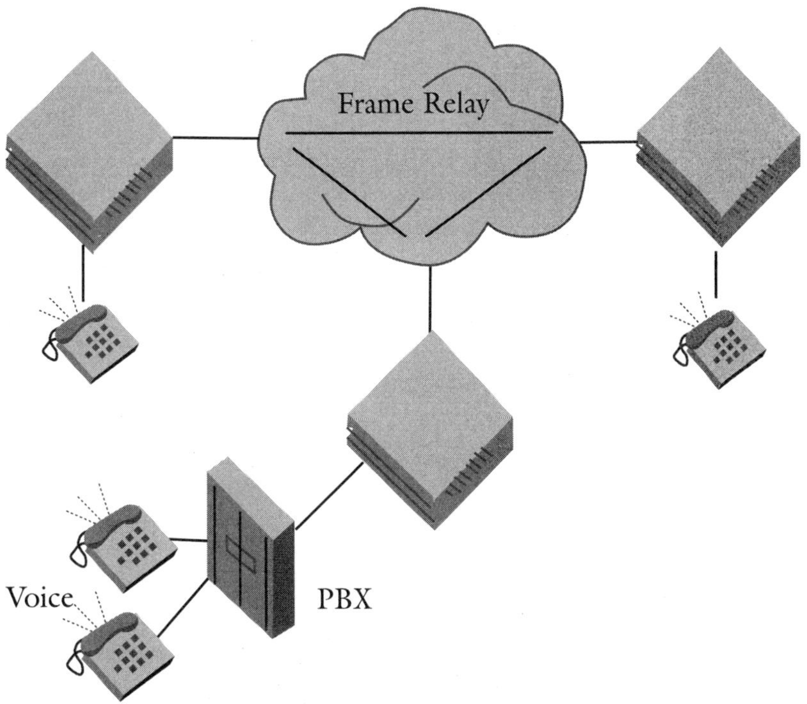

Multiple tandems should be avoided whenever possible regardless of the transport technology used. Both switching within the VFRAD and call processing can help alleviate some of the problems, and having both of these options available will let the network designer fine-tune the performance.

Equipment for Voice Over Frame Relay

The use of voice over a frame-relay infrastructure obviously involves equipment on the customer site, typically VFRAD as shown in Figure 3.11

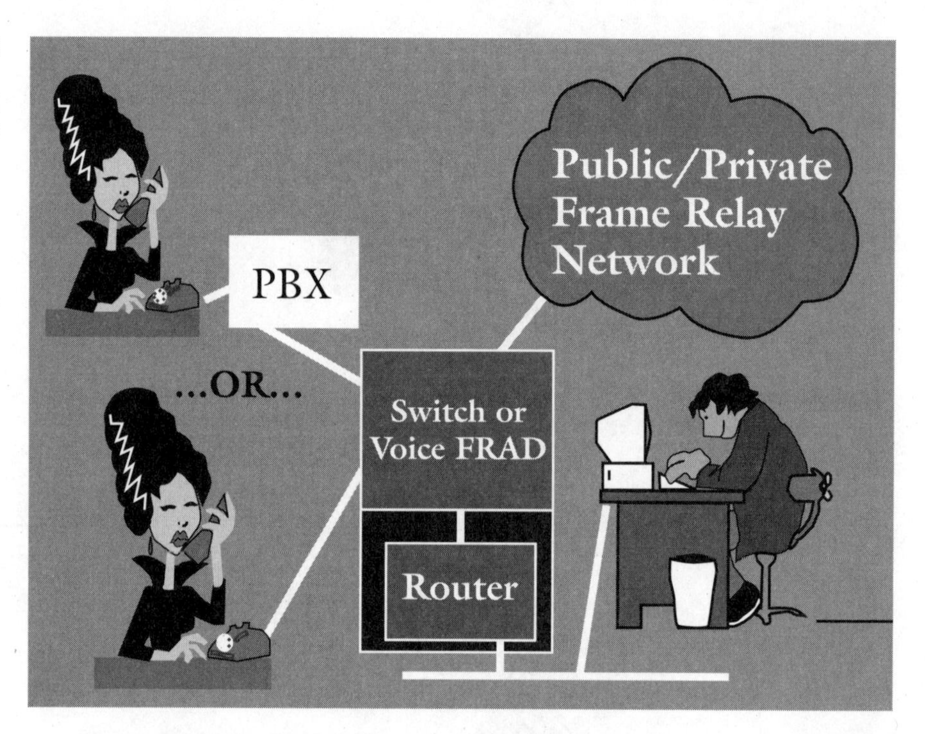

FIGURE 3.11 A typical VFRAD topology

A VFRAD is similar in function to a normal FRAD, but is specialized in that it handles—at a minimum—the voice packetization and compression process. Of course, some usual FRAD features are also included, like support for SNA and possibly some other serial protocols. Router support is provided either integrally or via external routers.

The VFRAD will typically be a point-to-point device with a single frame-relay interface. Nevertheless, when used in a networking environment, this single interface can support multiple logical connections to a number of sites on the frame-relay network by routing voice calls over several different virtual circuits. The VFRAD may also support at least one additional frame-relay interface. This is especially important for utilizing existing routers in conjunction with the VFRAD, allowing the VFRAD to perform frame switching. This additional interface could also allow for the connection of the VFRAD to a frame-relay network via two separate circuits and access points for additional reliability.

Data connections for existing routers and other frame relays accepting direct attachments can also support devices. Whether these capabilities are external or integral, the VFRAD must use priority and "fairness" algorithms to ensure that appropriate access to the network is available for all traffic types. This critical function is discussed in more detail in the next section.

The VFRAD typically supports a variety of voice interfaces. The most common is an analog interface, in which there is a direct conversion within the VFRAD from analog voice to compressed digital voice, with the compressed voice transported via frame relay. In this case, the voice trunks from the local PBX (or telephone) are attached directly to the VFRAD. The VFRAD, then, should support the most common analog voice functions, including a wide variety of 2-wire and 4-wire interfaces. If a telephone is directly attached, the VFRAD must supply battery and ringing functions.

A VFRAD also should support digital T1/E1 interfaces for headquarters and regional sites requiring a higher density of voice connections. These interfaces, typically from a PBX or PSTN, multiplex multiple conversations onto a single physical interface. This has the effect both of simplifying multiple connections and of reducing the requisite number of physical cables. It is also important to know how many conversations can be supported by the VFRAD since a T1/E1 interface can support 24/30 conversations.

Network Design Considerations

Frame relay provides statistically multiplexed access to many destinations over a single physical connection to the frame-relay network, as shown in Figure 3.12. To obtain the expected end-to-end performance, several important factors should be considered when designing the network.

Committed Information Rate (CIR)

CIR is the frame-relay parameter defining the minimum throughput that should be expected on a given virtual circuit. Design of TDM networks requires consideration of what portion of the bandwidth to allocate to each end-to-end connection. For a frame-relay network, consider how much CIR to allocate to each permanent virtual circuit (PVC). When the network is heavily loaded, each PVC will be able to carry at least the CIR allocated. When the network is lightly loaded, PVCs can carry substantially more than the allocated CIR.

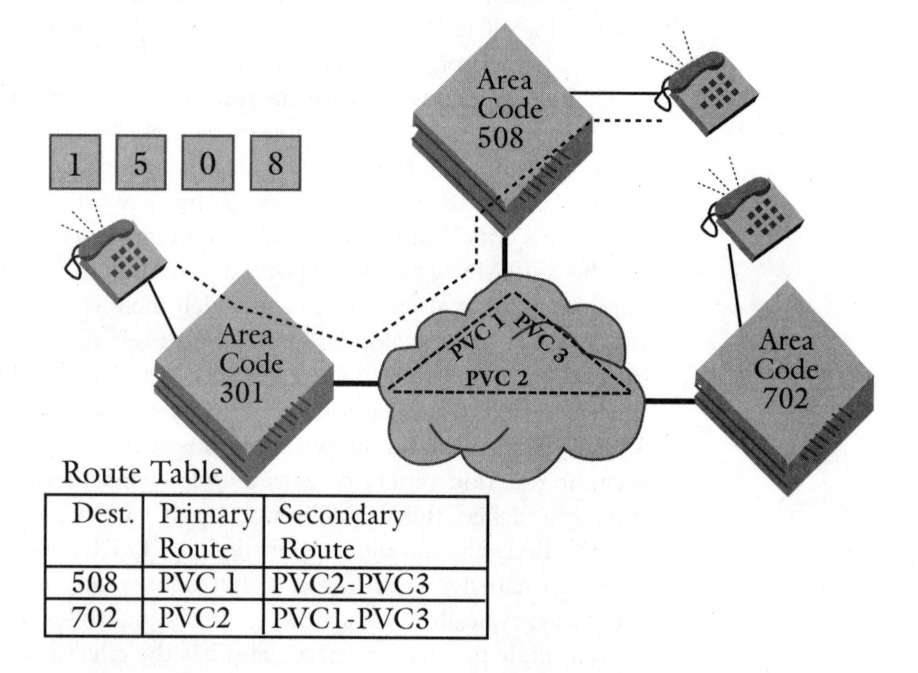

Dest.	Primary Route	Secondary Route
508	PVC 1	PVC2-PVC3
702	PVC2	PVC1-PVC3

FIGURE 3.12 A typical VFRAD topology

TDM and PVC? What are we talking about?

Time Division Multiplexer is a device that allows multiple conversations to share a single transmission facility, with each channel having access to a dedicated portion of the bandwidth.

Permanent Virtual Circuit is a virtual path through a network characterized by fixed endpoints defined by the network operator at service subscription. A single physical path may support multiple PVCs and SVCs. Compare with Switched Virtual Circuit.

Some public networks offer PVCs with zero CIR and provide very good performance because they design the network with enough capacity to prevent congestion except in case of a failure in the network. Others offer low CIRs and are so heavily loaded that very little traffic above the CIR can get through during peak periods. When traffic in excess of the subscribed CIR is sent into the network, the network tags the traffic as discard-eligible. During severe congestion the network will discard discard-eligible traffic, while CIR traffic is only discarded upon a failure in the network. Figure 3.13 provides a graphic view of this.

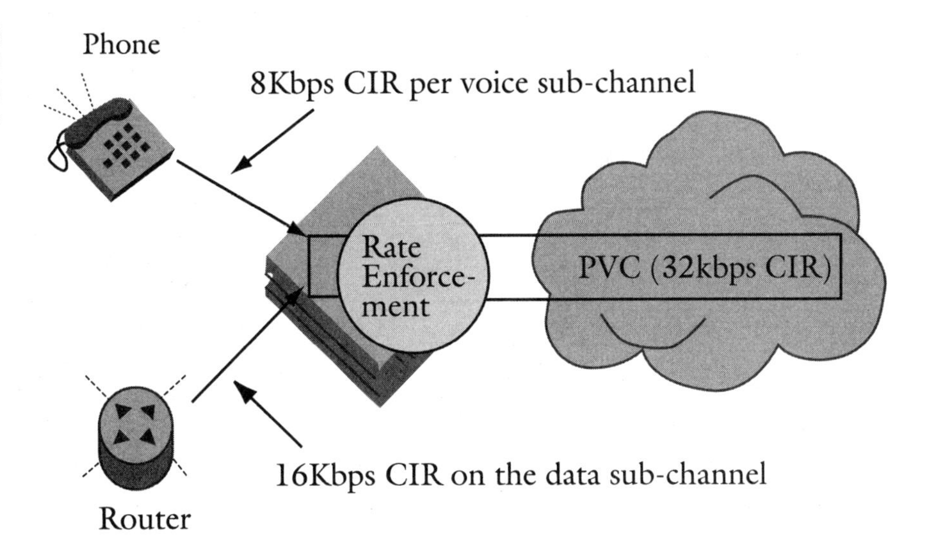

FIGURE 3.13 A typical VFRAD topology

Data traffic is likely to perform well when traffic is occasionally discarded, even if the traffic discarded is in a long burst. To recover, the user equipment re-sends the data traffic. Voice traffic can tolerate random discards (typically up to 10% can be discarded and good voice quality can still be maintained). However, if voice traffic were discarded in long bursts, entire spoken words could be missing from a conversation; this is not acceptable. To ensure long-burst discards do not occur, enough CIR should be purchased to ensure that voice traffic is not marked discard-eligible by the network.

CIR with Mixed Voice/Data Submultiplexing

To reduce recurring PVC costs, most VFRADs support submultiplexing. Submultiplexing allows many channels of voice and data traffic to be carried between two endpoints over single PVC as shown in Figure 3.14. If two 8Kbps voice channels are sharing one PVC, simply buy 16K of CIR for that PVC from the frame-relay service provider.

What happens when data are mixed with voice on one PVC? How much more CIR is needed? The answer to this can be quite complex.

If the source of the data is a high-speed router on a high-speed interface into the VFRAD, then it can quickly consume all the CIR on the PVC, which is being shared with the voice traffic; this is undesirable. To prevent this, the VFRAD must permit the network manager to configure CIR individually for each user of the PVC. When the data user (router) tries to consume more than the allotted CIR, the additional traffic is marked discard-eligible by the VFRAD and the network retains the remaining CIR for the voice users.

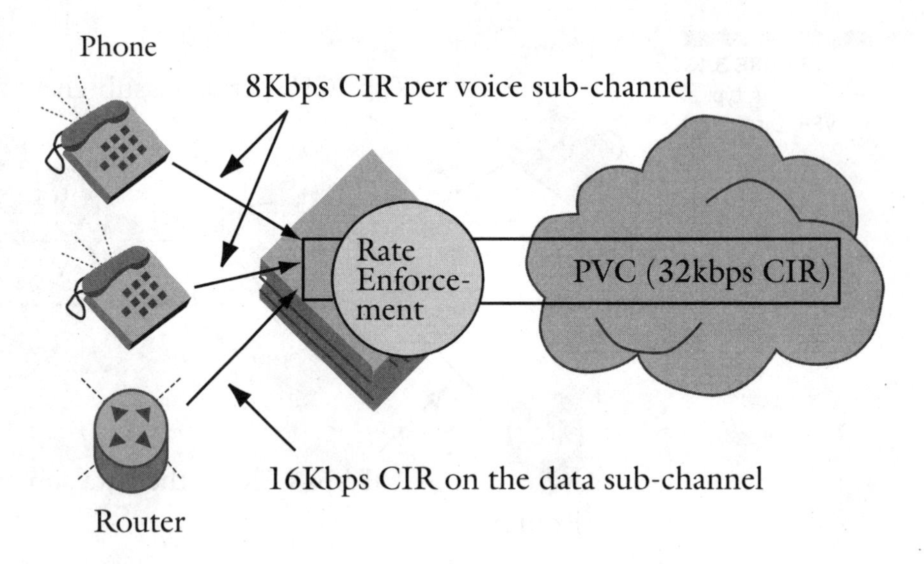

FIGURE 3.14 VFRADs' support of submultiplexing

High Speed Flooding and Traffic Shaping

While the VFRAD can enforce the rate at which CIR is consumed, it can also prevent an attached router from flooding excess frames over a high-speed interface, as shown in Figure 3.15. By configuring a restraining excess burst, the VFRAD can discard frames which exceed the limit. However, discarding traffic does not lead to optimum network performance.

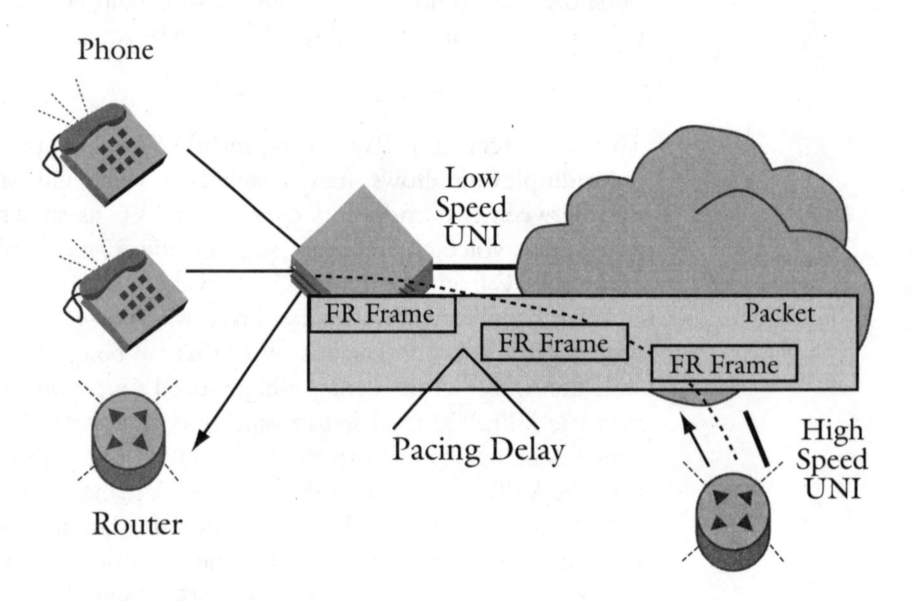

FIGURE 3.15 VFRAD not only enforces the rate of CIR consumed but prevents routers from flooding it

What about traffic flooding in from a distant router attached directly to the frame relay network? This excess flooding is an issue because the local frame-relay switch can get backed up with large amounts of traffic, which prevents voice frames from getting through in a timely fashion. The only good solution to the flooding issue is to configure the routers to pace their traffic per PVC at a rate compatible with the speed of the interfaces and relative to the peak load on the destination frame relay UNI.

A good example of routers configured to pace their traffic per PVC at compatible rates of speed and peak load of a destination frame-relay UNI is the traffic-shaping feature described in the Cisco IOS Software Release 11.2 Product Bulletin #487, section 3.2.2.

Another cause of flooding is very long data frames. VFRADs can control the effect of long data frames sent from attached routers by segmenting the frames into smaller pieces, sending them through the network, and then reassembling them at the destination VFRAD before delivering them to the destination router. When the destination is a router directly connected to the frame relay network the only solution is to configure the router to a maximum packet size compatible with the speed of the interfaces (slower interfaces require a smaller packet). In the near future there will be a Frame Relay Forum Data Fragmentation Implementation Agreement which will specify VFRAD to router fragmentation procedures. This will allow large packets to span multiple frames and will be configured per PVC.

Delay and Priority

Delay is critically important in voice applications and delay objectives can be easily obtained with a good network design. The end-to-end delay must be less than 250 milliseconds in each direction or else users will notice it. To attain this objective, the end-to-end delay through the VFRADs should be less than 100 milliseconds and the frame-relay network should deliver the traffic end-to-end in less than 150 milliseconds, as shown in Figure 3.16.

Delay through the network can vary from one moment to the next depending on network traffic load. VFRADs can compensate for normal delay variations. Under very heavy load conditions, with data users sending very long frames, delay variation can increase. Most frame-relay network switches offer PVC priority, which will allow traffic on one PVC to be sent while holding the lower-priority traffic for a short time. By configuring voice PVCs with a higher priority than data PVCs, delay variation can be substantially reduced. When PVC priority is not available, the other solutions are to place delay-sensitive traffic on a separate route through the network, or increase the network bandwidth so that it is not so heavily loaded.

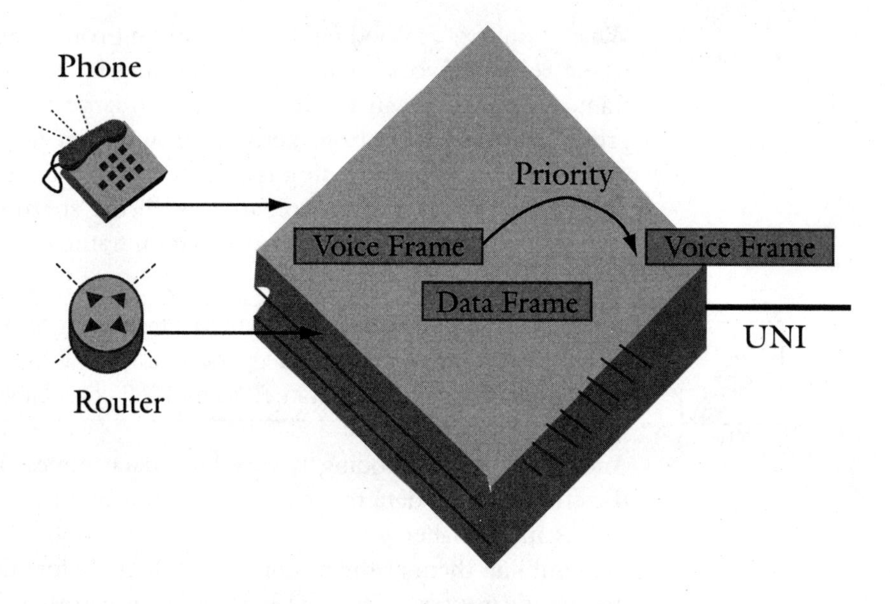

FIGURE 3.16 Configuring delay and priorities for VFRADs

Congestion Indication

The frame-relay congestion indicators, FECN and BECN, are sent from the network to the VFRAD when the traffic on a particular PVC encounters congestion. The VFRAD will respond in a variety of ways depending on where the traffic originated. Figure 3.17 depicts a layout of FECN and BECN indicators.

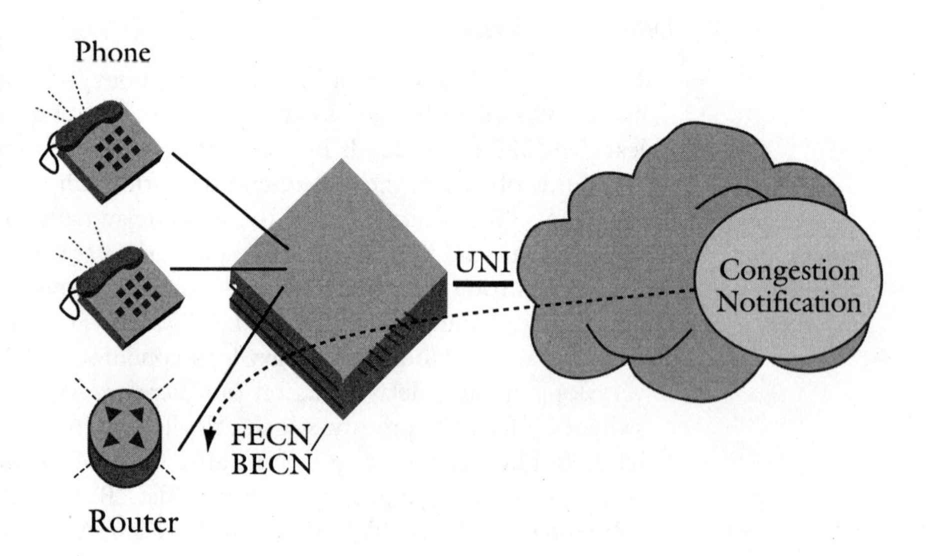

FIGURE 3.17 FECN and BECN frame-relay congestion indicators

If the data traffic was sent to the VFRAD from an attached router, the FECN and BECN will be passed to the router so that it can respond appropriately. If the data traffic originated from a VFRAD data port, the VFRAD will assert the appropriate flow control. When the CIR settings are configured as described above, voice traffic should never be a cause of congestion and therefore no action needs to be taken to reduce voice traffic.

Forward Explicit Congestion Notification (FECN) is a bit in the frame-relay header indicating that congestion may be present in the network for traffic traveling in the same direction as the direction of travel for the frame in which the bit is set.

Backwards Explicit Congestion Notification (BECN) is a bit in the frame-relay header indicating that congestion may be present in the network for traffic traveling in the direction opposite to the direction of travel for the frame in which the bit is set

Efficiency

Frames vary in length but the length of the header and trailer plus flag (the frame delimiter) is always five octets. Therefore, very long frames are very efficient and very short frames can be very inefficient. To reduce delay, voice frames are sent frequently and have a low average bit rate. As a result, voice frames can be quite short (typically 15 octets plus the five overhead octets at 8 Kbps) and inefficient. Figure 3.18 illustrates this concept.

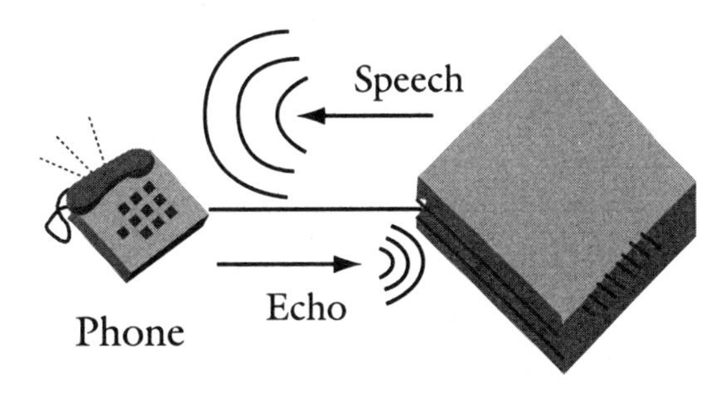

FIGURE 3.18 Very long frames are very efficient and very short frames can be very inefficient

Also, short frames place a burden on frame-relay switches, which have a limit on how many frames they can switch per second regardless of the frame length. To improve efficiency while slightly increasing delay, some VFRADs allow the network manager to configure (per user) multiple voice samples to be collected before placing them in a frame. This can dramatically add to the efficient use of the network bandwidth. Typically, each additional voice sample collected before sending a frame adds 15 milliseconds of delay. With three voice samples per frame, the end-to-end delay through the VFRAD should be around 100 milliseconds.

Echo Cancellation

Due to termination impedance mismatches between analog 2-wire circuits and 2- to 4-wire interface circuits at the called end of a network, voiceband echoes can be reflected back toward the calling end of the network, as shown in Figure 3.19. The voice-port echo cancellation feature, when enabled, can minimize these echo effects.

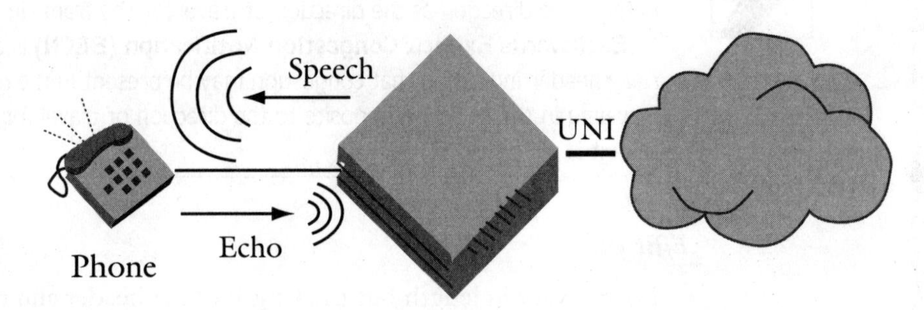

FIGURE 3.19 Voice-port echo cancellation features can minimize echo effects

Dialing Plan

Frame-relay PVCs interconnect PBXs in the same way they are used to interconnect LANs, as shown in Figure 3.20. By replacing leased lines with frame-relay PVCs, monthly costs are reduced and one PBX can connect to many other PBXs over one frame-relay network access line. How does the VFRAD know which PVC to use to get the voice call to the correct destination? It operates the same way the PBX would over multiple leased lines—it looks at the dialed digits and routes the call. The network manager configures the VFRAD with a dialing plan exactly as would be done for a PBX. This can be as simple as indicating the country code or area code for each destination, or can be more detailed if desired.

The VFRAD can also perform custom dial digit manipulation. For example, it can route a call based on a 10-digit number, and then outpulse only the last three digits of the number at the destination so that the destination PBX can ring the correct extension in an office. The dial-digit matching and substitution configuration is configured by the network manager and can be as simple or robust as desired.

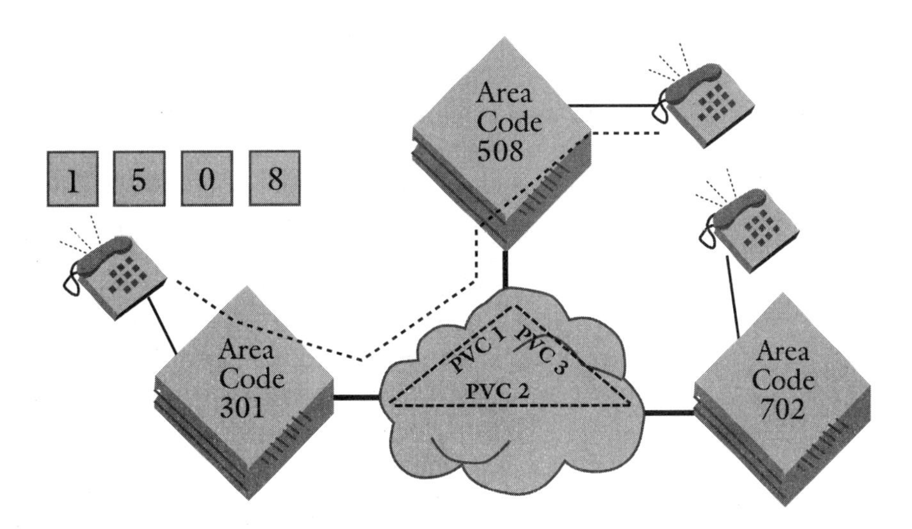

FIGURE 3.20 Frame relay PVCs interconnect PBXs in the same way they are used to interconnect LANs

Understanding Layer 3 Switching

Layer 3 refers to the network layer of the seven-level Open Systems Interconnect (OSI) model of networking. The network layer determines how data are transferred between computers, and address routing within and between individual networks. Conventional Ethernet switches work at Layer 2 (link level) of the OSI model. This requires external routers to transfer data among subnets. Integrating the routing function into the switch means users can implement switched networks without buying more routers. This helps reduce the cost of implementing a switched network, and reduces the overall cost of network ownership.

Routers cause bottlenecks in switched networks because they typically cannot transport more than 10,000 packets per second. Ethernet switches operate at up to 600,000 packets per second. Even though Layer 3 switching inherently alleviates router-caused bottlenecks, the actual improvement depends on how the routing is implemented. If it uses an off-the-shelf CPU, the same bottlenecks are likely to reappear because the switching speed will be limited by the CPU's processing time. That is because this approach requires the entire frame to go through the CPU.

Therefore, the pressure on networks is steadily increasing. Users are demanding more information faster and from increasingly distributed locations. At the same time, demanding new applications and skyrocketing Internet use are not only changing bandwidth requirements, they are also altering traditional traffic patterns.

A successful solution requires technologies that address performance issues at every level of the network, from the desktop to the telecommunications infrastructure. Fast Ethernet and emerging Gigabit Ethernet products, for example, offer high-bandwidth pipelines within the corporate Intranet to move data far more rapidly. Switching technology has also evolved to segment the traditional shared topology LAN, providing dedicated bandwidth where it is most needed.

Such developments have dramatically improved user access to information, but the resulting increase in data flow is creating new pressures at other levels of the network infrastructure. In particular, traditional backbone routers are being swamped as they try to direct traffic among high-bandwidth, switched networks.

Layer 3 switching offers a solution to this critical bottleneck. By integrating router functionality into the silicon within a switch, Layer 3 switching, such as Intel's illustrated in Figure 3.21, provides LAN-based routing at near-switching speeds. It is a significant innovation that can increase performance, while helping to reduce costs and complexity.

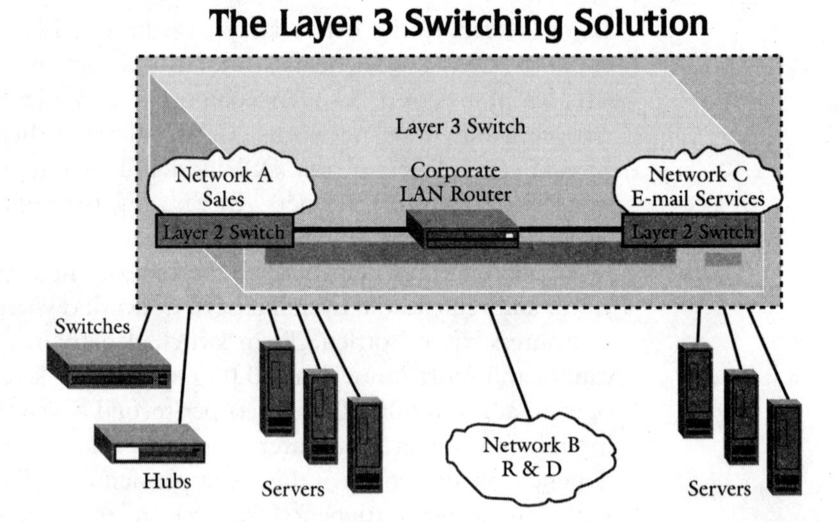

FIGURE 3.21 An Intel Layer 3 switch provides high-performance switching, plus LAN routing at near-switching speeds. (Source: Intel Corp.)

The hardware-based routing of a Layer 3 switch is much faster than traditional, software-based routing. Also, packets that need routing can travel across the backplane of the switch, providing yet another boost in performance. With the LAN router bottleneck removed, switched networks can take better advantage of available bandwidth. Desktop users get the high-speed network response they need, and the network is more stable and reliable.

A great feature of Intel's Layer 3 switching strategy outlined in Figure 3.21 is that it works to protect existing investments in network infrastructure, helping to ensure compatibility with current network components by using routing protocols well established as industry standards, unlike proprietary Layer 3 implementations. Even current routers remain useful, as they just move to the periphery of the network, where they can continue to handle WAN communications.

Why Layer 3 Switching?

Not so long ago, networks were small and flat, with simple peer-to-peer connections on a shared-media cable. Then these networks expanded and bridges were introduced to connect all the smaller networks into larger ones. However, as networks became busier and more complex, routers became the favored interconnection devices, because of their ability to provide segmentation and logical structure to the network. The result was the fully routed, hierarchical network, as shown in Figure 3.22; this is still a very common structure for intranetworks.

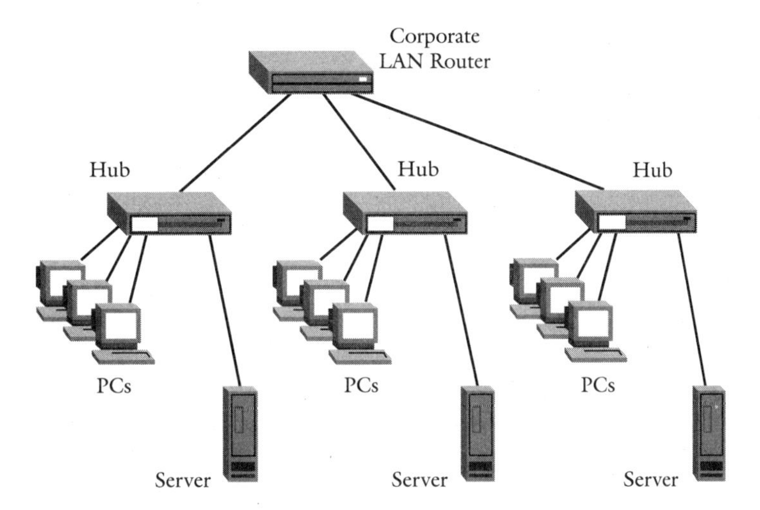

FIGURE 3.22 Traditional LAN routers segment the network and provide logical structure, but are slow, complicated and expensive (Source: Intel Corp.)

As these interconnected networks expanded and got busier, performance demands led to the development of switches (Layer 2 switches), as a fast, simple and cost-effective alternative to creating more routing layers within the Intranet, as shown in Figure 3.23. Corporate intranets became flatter again, and much faster—but they also became harder to control and less stable, because of their vulnerability to broadcast storms and redundant traffic.

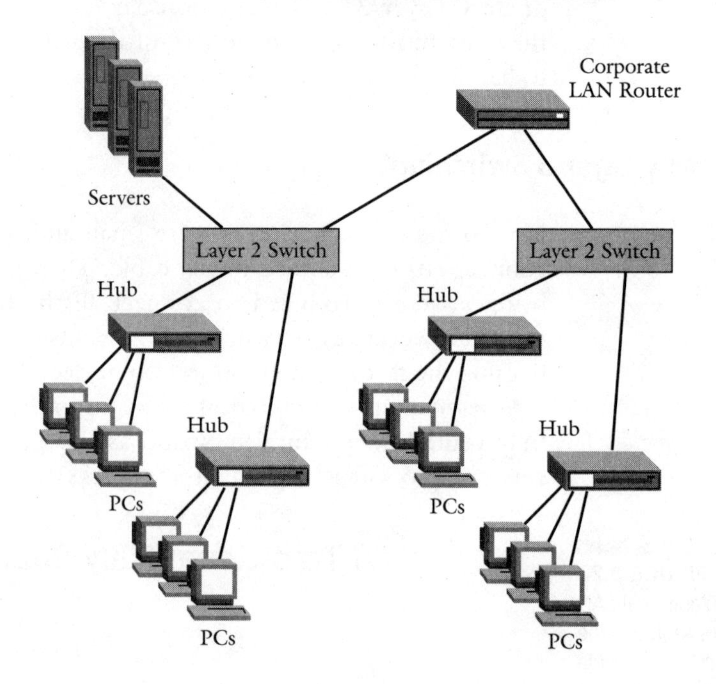

FIGURE 3.23 Standard switches are much faster than routers and provide dedicated bandwidth where needed (Source: Intel Corp.)

As the volume of Internet and intranet communications increased geometrically, more than 20 percent of network traffic crossing the boundaries of a local network became common. In many intranets, the amount of traffic directed across the limits of the local network is much greater than on a local network, as more users are connected to virtual private networks, extranets and so on. Non-local traffic is increasing beyond the capacity of LAN routers, and is putting huge pressure on WANs resulting in decreased network reliability and slower response times.

The countermeasure is usually a segmentation of the existing network into more and more switched segments. But the pressure on backbone routers has continued to rise, and a better solution was needed. Layer 3 switching was the answer. These "super-fast-packet-forwarders-with-some-routing-added" are aimed at relieving congestion in busy networks, such as the ones found in campus/building LANs, by offloading or replacing backbone routers that can no longer keep up. Layer 3 switches must work faster, scale better and be reasonably easy to deploy and manage. Layer 3 offers:

- ✔ **Enhanced Performance**—LAN routing at near-switching speeds eliminates router bottlenecks, while helping to improve support for high-bandwidth, multimedia applications (IP Multicast, VoIP)
- ✔ **Simplified Management**—Interoperates with existing network equipment and protocols, and is far easier than a router to install, configure and manage
- ✔ **Lower Cost**—Both acquisition and support costs are greatly reduced when a Layer 3 switch is used in place of a complicated and expensive router.

But What is Layer 3 Switching?

Layer 3 refers to the network layer in OSI's seven-layer model of networking, as shown in Figure 3.24. Layer 3 controls the routing of messages across different networks, as well as network flow and traffic management, and is the conventional dominion of routers.

FIGURE 3.24
Unlike a standard switch, which operates in the data-link layer of the OSI reference model, a Layer 3 switch also operates in the network layer to perform high-speed routing functions

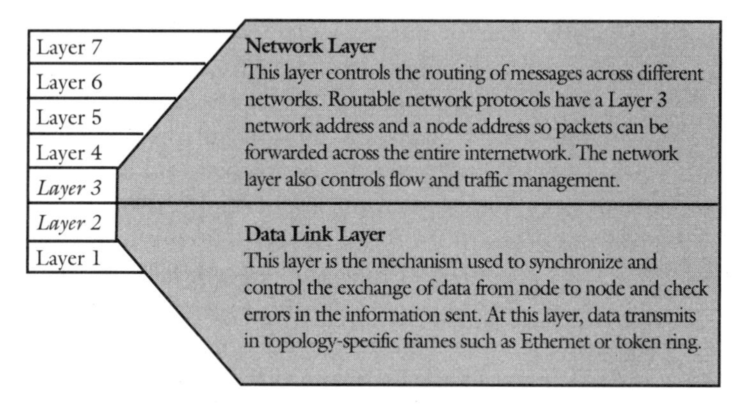

A typical switch operates at Level 2, called the data-link layer, which controls the flow of data between nodes. At this level, data transmits in topology-specific frames such as Ethernet. Layer 3 switches, in essence, operate at both levels, integrating the functionality usually associated with routers into the mechanism of a switch. Layer 3 switches are also known as ASCI-assisted routing, zero-hop routing, IP switching, NetFlow, tag switching, Fast IP, multiprotocol over ATM (MPOA) routing, route servers, and so on. But in all, Layer 3 switching products fall into one of two basic types of implementation categories:

- ✔ **Packet-by-packet layer 3 switches**—These full-blown routers examine every packet just as a router does and forward packets to their destinations. They run routing protocols such as OSPF, cache routing tables and understand the local-network topology. In function, there is little difference between routers and packet-by-packet Layer 3 switches. The contrast comes in price/performance. Packet-by-packet Layer 3 switches claim throughputs of a million packets per second (Mpps).
- ✔ **Cut-through layer 3 switches**—Cut-through describes a short-cut method of packet processing. Cut-through Layer 3 switches generally investigate the first packet or series of packets to determine destination. Once destination is understood, a connection is made and the flow is switched at Layer 2—delivering the low delay and high throughputs inherent in Layer 2 switching.

Both techniques deliver the high throughput benefits of a flat network without broadcast and security exposures. But each has pros and cons that should be considered:

- **Proven technology**—Packet-by-packet Layer 3 switches implement a known approach to internetworking, as opposed to cut-through techniques, which are very new
- **Performance**—Since packet-by-packet technique looks at every packet, it inherently suffers from more latency or delay in forwarding packets. Connection setup times and limitations of Layer 2 switch forwarding technology can slow cut-through switches as well
- **Distributed vs. centralized**—The packet-by-packet model is distributed, which means multiple Layer 3 switches in a large network; the centralized, minimized cut-through routing approach can become a bottleneck or single point of failure.
- **Interoperability vs. proprietary**—Cut-through technology lies somewhere between half-baked standards and quasi-open proprietary inventions, resulting in very little interoperability among vendors. Packet-by-packet Layer 3 switches can talk with any existing router in the network—and with other vendors' packet-by-packet switches.

TIP

For more information on Layer 3 switching, in particular Intel's products, check the company's Web site: **www.intel.com/network/**.

In summary, as networking devices have gotten more complicated, the marketing language used to describe their capabilities has gotten more fuzzy and general. No two switches are identical, and there are many technologies to chose from that can deliver fast network connectivity, essential to VoIP applications. Gigabit Ethernet is another concept worth considering, not for the understanding of the technology, but so that it can be differentiated from the resources offered by Layer 3 switches.

Introduction to Gigabit Ethernet

As PCs become more powerful, applications demand more bandwidth, and users access new media formats such as multimedia, video, intranets and the Internet, the ability of current network bandwidth to handle growing user needs becomes a central issue, as Figure 3.25 shows.

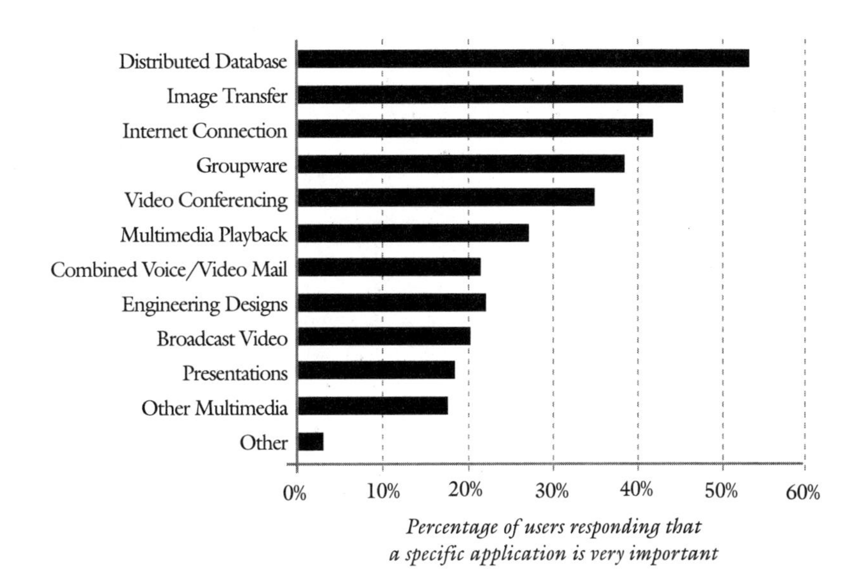

FIGURE 3.25 Existing and emerging applications are driving the need for more bandwidth (Source: Infonetics March 1996)

At the same time, new higher-bandwidth solutions must be backward-compatible with existing technologies, to protect large investments in network infrastructure. For this reason, and others, Gigabit Ethernet is emerging as an industry-standard solution for high-speed local area networking. Intel is committed to high-performance networking with a considerable customer focus. Intel wants to provide users with better, faster, more affordable access to computing power and information.

Intel plays a leadership role in network technology, bringing unique strengths to the networking arena, as it has an in-depth knowledge and expertise in the PC system, including CPU development, PCI bus design, system architecture, and LAN connectivity hardware and management software. Also, as a founding member of the Fast Ethernet Alliance, which expanded the capacity of Ethernet tenfold, Intel also has the design and manufacturing infrastructure necessary to deliver top-performing products and value.

Fundamentals of Gigabit Ethernet

Gigabit Ethernet is an extension of the 10Mbps (10BASE-T) Ethernet and 100Mbps (100BASE-T) Fast Ethernet standards for network connectivity, as depicted in Figure 3.26. The IEEE has given approval to the Gigabit Ethernet project as the IEEE 802.3z Task Force, and the specification was expected to be complete in early 1998. There have been more than 200 individuals representing more than 50 companies involved in the specification activities to date.

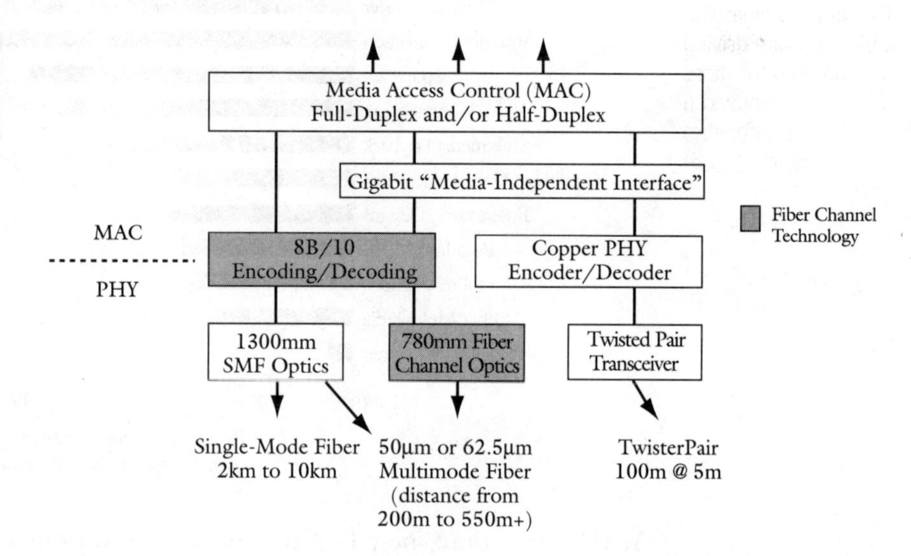

FIGURE 3.26 Fundamentals of the Gigabit Ethernet technology

Fully compatible with the huge installed base of Ethernet and Fast Ethernet nodes, Gigabit Ethernet will employ all of these same specifications as the original Ethernet specification (defined by the frame format and support for Carrier Sense Multiple Access with Collision Detection protocol, full duplex, flow control, and management objects as defined by the IEEE 802.3 standard). Thus, Gigabit Ethernet uses the same Ethernet technology readily available, but is ten times faster than Fast Ethernet and 100 times faster than Ethernet.

Gigabit Ethernet Benefits

Gigabit Ethernet offers enhanced benefits that enables fast optical fiber connection at the physical layer of the network. It provides a tenfold increase in MAC (Media Access Control)-layer data rates to support video conferencing (and VoIP), complex imaging and other data-intensive applications. Also, as just mentioned, Gigabit Ethernet has the advantage of being compatible with the most popular networking architecture, Ethernet. Since its introduction in the early 1980s, Ethernet deployment has been rapid, quickly overshadowing networking connection choices such as token ring and ATM.

There is no need to purchase additional protocol stacks or invest in new middleware when deploying Gigabit Ethernet. According to IDC research projections, more than 80 percent of installed connections were Ethernet back in 1996. IDC predicts that Ethernet will continue to prevail, and it is expected to continue its growth beyond 1998, particularly as this compatible and scalable standard moves to gigabit speeds. Figure 3.27 show the increased number of faster network interface cards (NIC) versus standard 10Mbps sold.

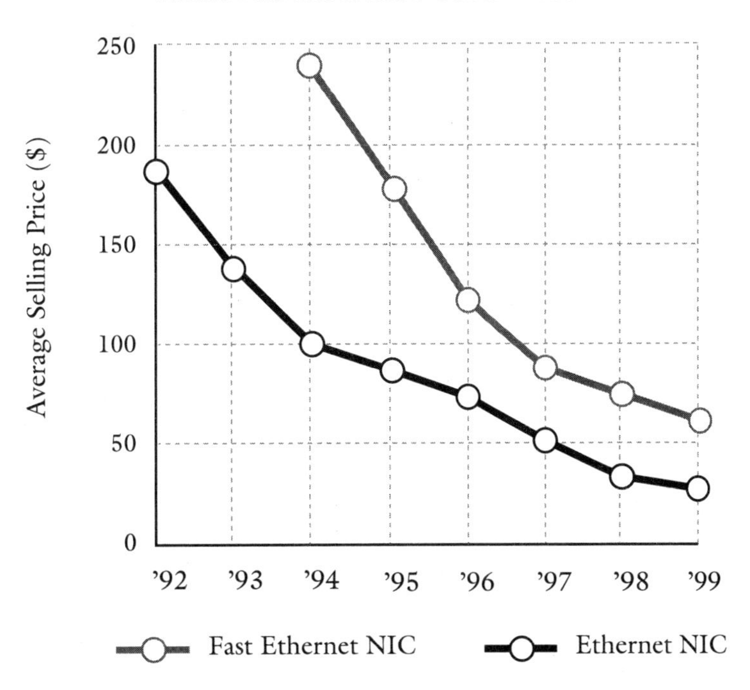

FIGURE 3.27 Ethernet and Fast Ethernet NICs have shown steady cost reductions over time. Similar trends are anticipated for Gigabit Ethernet products. (Source: Dell Oro Group)

As IT departments adopt Fast Ethernet, and eventually Gigabit Ethernet, to enhance network performance to support robust desktop needs, they will see:

- ✔ Increased network performance levels, including traffic localization and high-speed cross-segment movement
- ✔ Increased network scalability—it will be easier to add and manage more users and "hungrier" applications
- ✔ Decreased overall costs over time.

The proliferation of Intel Pentium, Pentium Pro and Pentium II processor-based desktops in corporate networks, combined with new bandwidth-intensive operating systems and applications, has already influenced many LAN decision makers to migrate to Fast Ethernet. This anticipated growth is based on the fact that, unlike FDDI (Fiber Distributed Data Interface) and ATM, Gigabit Ethernet addresses the bandwidth dilemma without requiring costly protocol changes.

How Gigabit Ethernet Measures Up Against Other High-speed Solutions

ATM is among some of the alternatives to enhance Ethernet performance. Although adoption of Gigabit Ethernet does not exclude ATM as a solution within an overall LAN/WAN architecture, according to IDC #12382, Gigabit Ethernet is rapidly emerging as the preferred solution, as shown in Figure 3.28.

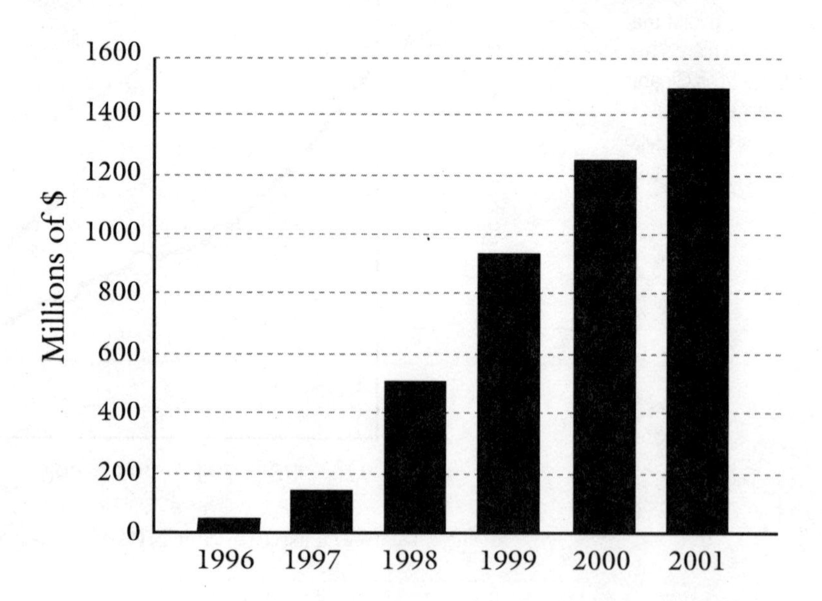

FIGURE 3.28 Predicted growth of Gigabit Ethernet products. (Source: IDC #12382, Nov. 96)

The Gigabit Ethernet Alliance conducted a recent study and found out that the majority of respondents planned to evaluate or deploy Gigabit Ethernet technology within the next six months to a year, and that Gigabit Ethernet is the preferred solution for switch-to-switch connections, catching up to ATM as the technology of choice for LAN backbones. Figure 3.29 illustrates this finding.

FIGURE 3.29 Gigabit Ethernet is widely popular, catching up with ATM while offering a more comprehensive solution

Capability	Gigabit Ethernet	ATM
IP Compatibility	Yes	Requires RFC1577 or PNNI implementation
Ethernet Packets	Yes	Requires LANE routing from cells to packets
Handle Multimedia	Yes	Yes, but applications need to change
Quality of Service	Yes, with RSVP and 802.1Q	Yes, with SVCs

Gigabit Ethernet's wider bandwidths help improve QoS, regulating the timing of latency periods to minimize jittery video and audio delays. In the past, ATM was the only reliable way to achieve any kind of QoS. Today, Gigabit Ethernet is rapidly closing the gap, and with considerably more economy, backward compatibility and interoperability with other technologies. Thus, very likely, ATM will remain at the WAN level of interconnectivity. It is unlikely that ATM will ever move down to the workgroup or desktop, because it would require a complete change of network interface hardware, software and management protocols. Figure 3.30 shows how Gigabit Ethernet delivers many of the benefits originally expected from ATM, but is much easier to implement, as well as cost-effective.

When it comes to Gigabit Ethernet, watch Intel, which has established itself as a leader in the transition to Fast Ethernet, with its family of Fast Ethernet desktop, server and mobile adapters, print servers, hubs and switches. The PCI bus for Intel architecture PCs and servers is tailor-made for today's power users. A 32-bit PCI implementation already pumps out data in the multi-hundred megabits range. In the future, a 64-bit PCI bus will easily handle Gigabit Ethernet throughput at the desktop.

Also, Intel's ongoing relationships with key industry leaders, such as Cisco and Microsoft, positions Intel's commitment to extending and supporting industry standards, as this cooperation will assure compatibility with Gigabit Ethernet products that emerge from other vendors.

FIGURE 3.30 Gigabit Ethernet can deliver many of the ATM features, at a fraction of the cost, and with easier integration

	ATM	Ethernet & Fast Ethernet	Gigabit Ethernet
End-to-End	✔		✔
Scalable	✔		✔
Connection-Oriented	✔		✔
Quality of Service	✔		✔
Low Cost		✔	✔
Interoperability		✔	✔
Standards		✔	✔
Software		✔	✔
Ease of Integration		✔	✔

What's Next

This chapter introduced the basic concepts of voice over IP (VoIP) and the technologies surrounding it. It discussed the H.323 standard as well as other standards and technologies such as audio codecs, IP over ATM, voice over ATM, the emulation of traditional T1/E1 Trunks, IP over SONET and voice over SONET, and IP and voice over frame relay. It introduced the concepts of Layer 3 switching and Gigabit Ethernet and their role in VoIP.

The next chapter deepens the discussion of IP multicasting by considering multicasting in workgroups, some of its capabilities on hosts and routers, as well as usage and implementation, especially with VoIP.

CHAPTER 4

More on IP Multicasting

Multicast techniques span many areas of networking, including video and teleconferencing, multimedia presentations, news distribution, and remote live broadcasts such as those from space. IP multicast can run over just about any network infrastructure including Ethernet, ATM, frame relay, SMDS, and satellite. For multicasting to work, multicast-aware TCP/IP stacks need to be installed on all the participating machines.

As discussed in Chapter 2, multicasting is connectionless, which means that a multicast datagram is neither guaranteed to reach all members of the group nor guaranteed to arrive in the same order as it was sent. The protocol delivers a multicast datagram to the destination group members on a best-effort basis.

A best-effort basis can introduce latency and variability of delay in end-to-end paths. However, multicast, applications require control over the QoS they receive. The level of security, bandwidth, delay, jitter, error rates, cost, and compression are some of the parameters that differentiate the network services that QoS provides. The RSVP is a key protocol that makes QoS possible.

The protocols and algorithms used by multicasting applications are diverse and complex. Multicasting can be implemented at several layers of the OSI model, on different media, and with different protocols.

TIP

For more information on IP Multicasting, please refer to my book, co-authored with Kitty Niles, IP Multicasting: Concepts and Applications (McGraw-Hill).

IP multicast is an extension of IP. The IETF-recommended standard, RFC 1112, defines extensions to the IP. A relatively new feature of the IP Multicast is a protocol for transmitting IP datagrams from one source to many destinations in a LAN or WAN. Groups of receivers participate in multicast sessions. With IP multicast, applications send one copy of the information to a group address. The information reaches all the recipients who want to receive it. Multicast technology addresses packets to a group of receivers rather than to a single receiver; and depends on the network to forward the packets only to the networks that need to receive them. Multicast-enabled nodes that run the TCP/IP suite of protocols can receive multicast messages.

An Overview

Multicasting, as shown in Figure 4.1, is a push technology where a server sends data to a client without the client requesting it. In pull technology a client requests data from a server or from another computer. E-mail and PointCast are examples of push technology while the Web is based on pull technology.

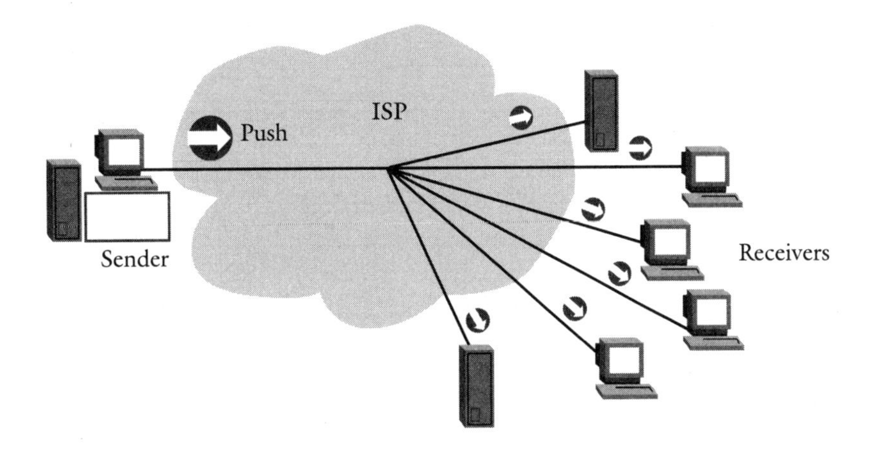

FIGURE 4.1 An example of multicasting to multiple hosts on the Internet

Standards-based IP multicasting supports thousands of users simultaneously without substantially affecting bandwidth requirements. In addition, IP multicast routing protocols provide efficient delivery of datagrams from one source to any number of destinations throughout a large, heterogeneous network such as the Internet. If the network hardware supports multicast, then packets destined for multiple recipients can be sent as a single packet.

There are three fundamental types of IPv4 addresses: unicast, broadcast, and multicast. A unicast address is designed to transmit a packet to a single destination. A broadcast address is used to send a datagram to an entire subnetwork. A multicast address is designed to enable the delivery of datagrams to a set of hosts configured as members of a multicast group in various scattered subnetworks.

Types of Transmission

Traditional transmission methods, unicasting and broadcasting, differ in numerous ways from IP multicasting. Mapping IP multicasting to LAN multicasting methods involves three separate operations: multicast address resolution to LAN multicast addresses, copying and forwarding of messages, and group membership registration.

A unicast address is designed to transmit a packet to a single destination. A unicast transmission is inherently point-to-point, as shown in Figure 4.2.

If a node wants to send the same information to many destinations, it must send copies of the same data to each recipient in turn. The same information must be carried over the network multiple times. Unicast avoids sending the data to networks where there are no stations that need it but does use up network bandwidth and resources. In addition, the node needs to generate separate identical data streams for each recipient. This is very inefficient, using up processing power and memory.

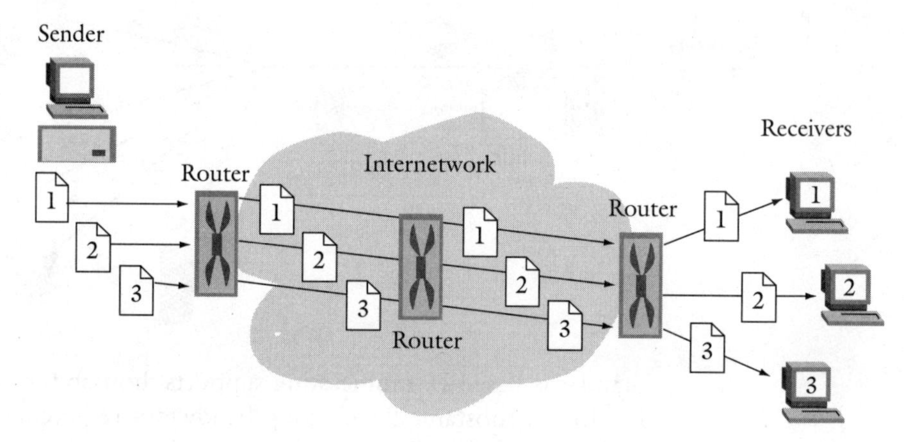

FIGURE 4.2
A basic unicast point-to-point transmission

A broadcast allows one station on the network to talk simultaneously to all devices contained in the same broadcast domain, or subnet. Routers and switches forward broadcast but in doing so use bandwidth and have no way of knowing if any of the nodes on the other network want the broadcast data. Broadcasting does not consume the *sender*'s resources any more than single unicasting, but inefficiently consumes ***network*** resources.

Some protocols use broadcasting to discover resources from the network. To prevent broadcast messages from flooding the network, system administrators may configure routers to pass or block broadcast on any particular route. Many data communication networks restrict broadcasting to one physical or logical segment of the network.

A multicast address enables the delivery of a singe data stream to a set of hosts configured as members of a multicast group in various scattered subnetworks. Therefore, multicasting is a method to reach several recipients by one transmission, as shown in Figure 4.3. Other nodes filter out multicast packets at the hardware level. Multicasting is the process of sending to a self-selected group of recipients that is often substantially less than the full population of recipients. Each recipient must be defined separately and it must be possible to control which recipients receive data. Multicast dynamic groups of recipients can be created and removed very quickly.

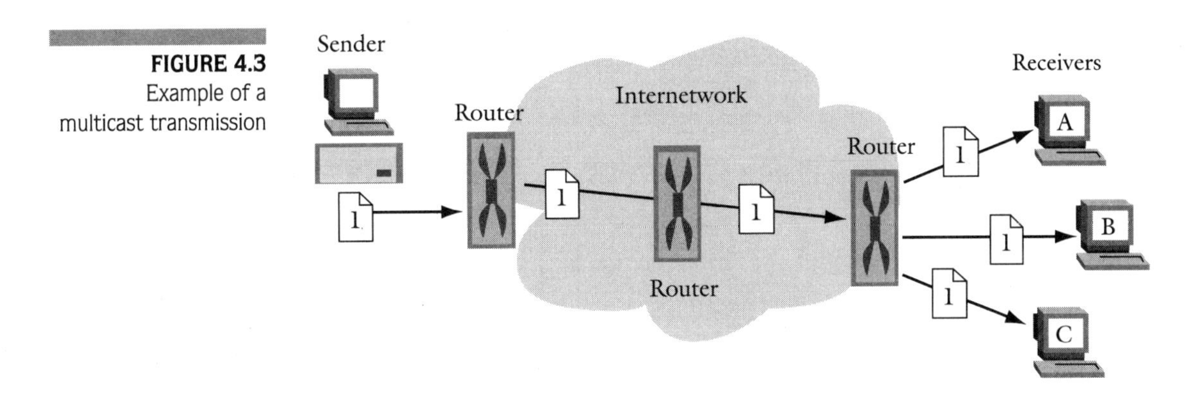

FIGURE 4.3 Example of a multicast transmission

Multicast Address Translation

A multicast address enables the delivery of datagrams to a set of hosts configured as members of a multicast group in various scattered subnetworks. A Class D address is a multicast address and identifies the group of machines or interfaces that represent a multicast group. For example, a Class D address could identify all the interfaces attached to IP network routers. All logical holders of a particular Class D address receive packets sent to that address.

The destination-address field of the IP header in a multicast IP packet contains a Class D group address instead of a Class A, B, or C IP address. A Class D address is an IP address and has the format 224.0.0.0 - 239.255.255.255.

In Class D addressing the lower 256 entries of the address range are reserved for administrative functions and system-level routing chores. The middle range is for use by multicast end-user applications within groups, intranets, and the Internet. The upper range of the Class D address set is reserved for locally administered or site-specific multicast applications, as shown in Figure 4.4.

When a local router on a subnet receives a Layer 3 multicast packet, it can map the IP multicast address to a Layer 2 multicast address, such as an Ethernet MAC address. The receiving host's LAN interface hardware can efficiently read this Layer 2 address. Layer 2 LAN protocols typically reserve portions of their address space for broadcast and multicast frames, for example, the Ethernet broadcast address FF-FF-FF-FF-FF-FF.

Address translation from the IP (Layer 3) address to the Layer 2 address occurs by directly mapping the IP address into an Ethernet MAC address. This is accomplished by dropping the low-order 23 bits of the IP multicast address into the low-order 23 bits of the Ethernet multicast address.

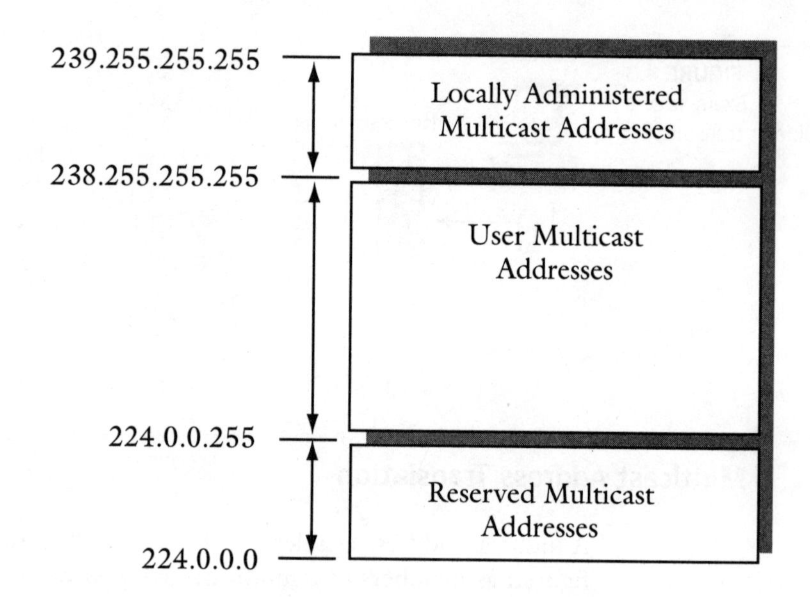

FIGURE 4.4 Class D multicast addressing

Multicasting and Routing

Sending the information just once to multiple users can create big savings in bandwidth. Copies of the message are made only when paths diverge at a router such as when the message is supposed to be passed on to another router as well as to a workstation attached to the current router.

Multicast-enabled routers forward a multicast to a network only if there are multicast receivers on that network. Host machines use the Internet Group Management Protocol (IGMP) to inform a multicast-aware router of any multicast sessions in which they want to participate. If all members of a multicast group on a particular network segment leave the group, the router ceases to forward multicast data to that segment.

IP Multicasting Uses and Benefits

In today's business environment large amounts of information or products need to be delivered to multiple sites in real time. At the same time business and research need to retrieve large amounts of passive or static data on a daily basis. New data communication networks have created much more capacity. The increased capacity has created new possibilities to develop innovative services. These new services have created a need for new transmission methods including the underlying communications infrastructure.

Currently the majority of Internet applications rely on point-to-point transmission that has traditionally been limited to local area network applications. Business is now very frequently a global affair.

IP multicasting forces the network to do packet replication, which conserves bandwidth. IP multicast is a good alternative to unicast transmissions when a company needs to deliver applications to multiple hosts at the same time. It is important to note that the applications for IP multicast are not solely limited to the Internet. Multicast IP can also play an important role in large distributed commercial networks.

Network Load Reduction

IP multicast can reduce the load on the network. If, for instance, an application needs periodically to transmit packets to several hundred hosts within the company, IP multicasting can be the solution. Periodic unicast transmission of these packets would require many of the packets to traverse the same links. Multicast transmission of those same packets would require only a single packet transmission by the source. This transmission is then replicated at forks in the multicast delivery tree.

Broadcast transmission is not an effective solution for this type of application since it affects the CPU performance of each end station that sees the packet and it wastes bandwidth.

Internet multicasting is the only standards-based solution that can support thousands of users simultaneously without affecting bandwidth requirements. In response to business needs, some, but not all of today's ISPs support IP multicasting.

Also, IP multicasting enables the distribution of internal corporate data to large numbers of users. A company with a chain of stores could use multicast to send pricing information to cash registers company-wide. This preserves bandwidth locally and across the network. Multicasting multimedia data across the Internet and intranets to multiple users is an excellent way to preserve bandwidth. For example, providers of live real-time information feeds can use Internet multicasting to deliver their content. Such providers in the United States include CNN, ESPN, and Home Shopping Network.

VoIP and Video Conferencing

Interactive video conferencing and other VoIP applications using multicasting though the Internet, Intranet or extranet are becoming an economical alternative to expensive ISDN-based solutions. IP multicasting is providing:

- ✔ Productivity because it decreased travel time
- ✔ Large cost benefits from savings in travel expenses
- ✔ The ability for a large percentage of the employees, rather than just a few, to" stay current" by attending video conferences.

Rather than using expensive ISDN-based equipment or services, companies could use multicast-capable computers connected through the Internet, Intranet, or extranet to address their video conferencing needs

Another important part of today's business computing environment is the sharing of information through databases rather than just traditional e-mail. This can be accomplished through groupware applications such Lotus Notes, Microsoft Exchange, and Corel GroupWise. The information can be in databases scattered throughout the company or the world. Keeping the databases synchronized in all locations and updated with the most recent information is important. All groupware users must have the same information at the same time.

The use of IP multicasting in this instance makes it possible to use one transmission to send changes instantly to all the databases. All databases would instantly contain the same information. In the book *IP Multicasting: Concepts and Applications*, the example was given of a corporate user in Brazil, another in Finland, and yet another in the United States who all read the same information at the same time as shown in Figure 4.5

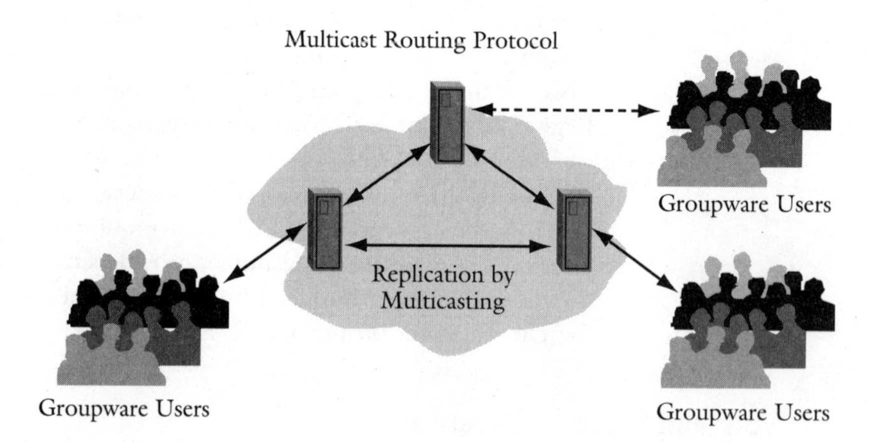

FIGURE 4.5 Sharing information through a groupware network

A Word About the Multicast Backbone

The specification for IP multicasting was published in 1989, but its use has been limited. The number of routers that support multicasting on the Internet has not been large, but this is changing slowly. Routers on the Internet tend to be replaced by multicast-capable ones only when a new router is required to replace an old one.

In the meantime, researchers wanted a resource and test bed for testing multicasting protocols and applications. They also wanted a way to enable the deployment of multicast applications without having to wait for multicast-enabled devices to be installed through the Internet. They therefore developed the Internet Multicast Backbone (MBone). The MBone supports routing multicast packets without disturbing or altering other Internet traffic and has been in existence for about six years.

The MBone is an experimental, cooperative volunteer effort spanning several continents. It is an interconnected set of subnetworks and routers that support the delivery of IP multicast traffic. A virtual network layered on top of the Internet, the MBone bypasses multicast-unaware routers on the Internet using tunnels. To this end, the Distance Vector Multicast Routing Protocol (DVMRP), described in RFC 1075, has been used to build the MBone by building tunnels between DVMRP-capable machines. The endpoints of the tunnels are entered manually in routing tables and administrated in the MBone. Figure 4.6 describes tunneling.

FIGURE 4.6
Mbone tunneling

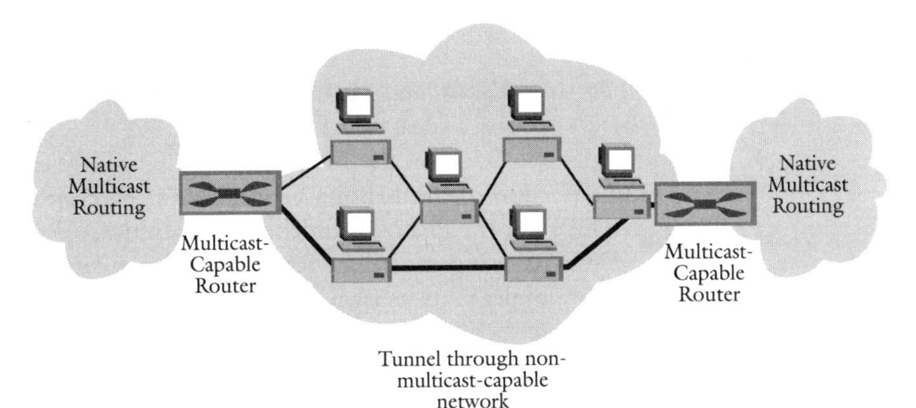

MBone is constructed with tunnels across networks that do not support multicast routing. The tunnels allow multicast traffic to pass through the non-multicast-capable parts of the Internet. The MBone mostly uses encapsulated tunnels between multicast-capable islands of the Internet to move multicast data. An IP multicast packet traversing an encapsulated tunnel is characterized by its IP source and destination addresses being the IP addresses of the tunnel endpoint multicast routers. Figure 4.7 is a simplified representation of the MBone showing the concept of multicast-capable islands.

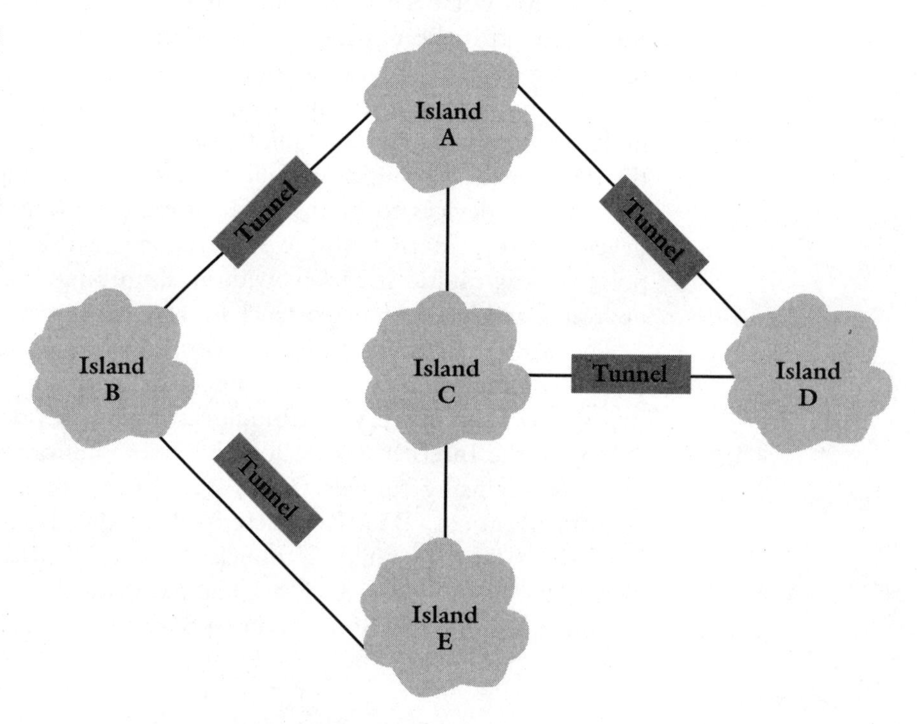

FIGURE 4.7
Internet MBone

Since the MBone and the Internet have different topologies, multicast routers execute a separate routing protocol to decide how to forward multicast packets. Most of the MBone routers currently use DVMRP. There are some portion of the MBone however, that execute either Multicast OSPF (MOSPF) or the Protocol-Independent Multicast (PIM) routing protocols.

DVMRP is a protocol for routing multicast datagrams through an Internet and implements its own unicast routing protocol in order to determine which interface leads back to the source of the data stream. This unicast routing protocol is based purely on hop counts. Because of this, the path the multicast traffic follows may not be the same as the path the unicast traffic follows.

Nowadays the IEETF, NASA, and research groups world wide use the MBone for research and testing of multicast protocols and services, for multicast multimedia recordings of meetings and live space events across the Internet, and for desktop conferencing. Even live concert performances have been multicast over the MBone. The number of sites participating in the MBone has grown rapidly. As an experimental and volunteer effort, the MBone has limited its use in commercial environments.

The Capabilities of Multicasting

An IP multicast-capable network forwards multicast packets according to the group address of the packet. Network routers that support multicast keep track of which parts of the network have multicast hosts joined to particular groups. Routers forward multicast packets only to subnetworks with IP multicast-capable hosts joined to the particular group. Multicast-capable hosts and routes have certain requirements.

RFC 1112 describes the IP multicast extensions to the standard IP protocol. Three levels of conformance to this standard exist. A Level 0 host has no support for IP multicasting and multicast activity has no effect on them.

To provide Level 1 multicasting, a host IP implementation must support the transmission of multicast IP datagrams. At Level 1, hosts can only send multicast datagrams. Level 1 allows a host to partake of some multicast-based services, such as resource location or status reporting. A Level 1 host cannot receive multicast datagrams.

To provide Level 2 multicasting, a host must also support the reception of multicast IP datagrams. At Level 2 host have full support for IP multicasting. Hosts can join and leave multicast groups and receive multicast datagrams sent to group addresses. Level 2 requires implementation of IGMP and extension of the IP and local network service interfaces within the host.

The most important part of host IP multicast support implementation is Internet Group Management Protocol (IGMP).[1] Both IGMP and ICMP reside in the IP layer. IGMP's function is to keep neighboring multicast routers informed of the host group memberships present on a particular local network and to provide the mechanisms by which hosts and routers can join and leave IP multicast groups. The mapping of IP addresses to local network addresses is considered to be the responsibility of local network modules.

IGMP uses the reserved multicast group address 224.0.0.1 to communicate with local routers. Called the "all-hosts group", this multicast group addresses all hosts in the LAN. It is through this channel that IP multicast routers learn if any hosts are joined to a multicast group in this particular LAN. Routers send IGMP queries to this address at the LAN and hosts respond by telling which groups they want to join.

1 The IGMP protocol has been updated and is available in RFC 2236 by W. Fenner, "Internet Group Management Protocol, Version 2," November 1997.

The mapping between a Class D IP address and an Ethernet MAC-layer multicast address is obtained by placing the low-order 23 bits of the Class D address into the low-order 23 bits of IANA's reserved MAC-layer multicast address block. Mapping from the Class D group address to the MAC address is not one-to-one, because the high 5 bits of the Class D group address are discarded.

IP multicasting is an extension of link-layer multicast to IP Internets. Using IP multicasts, a single datagram can be addressed to multiple hosts without sending it to all hosts. An IP datagram sent to the group is delivered to each group member with the same best-effort delivery as that provided for unicast IP traffic.

Forwarding of IP multicast datagrams is accomplished either through static routing information or through a multicast routing protocol. Devices that forward IP multicast datagrams are called multicast routers. They may or may not also forward IP unicasts. Routers forward multicast datagrams on the basis of both their source and destination addresses. An IP multicast packet traversing an encapsulated tunnel such as in the MBone is characterized by its IP source and destination addresses being the IP addresses of the tunnel endpoint multicast routers.

In response to the growing need for multimedia applications and real-time data distribution, network-layer multicast services are being built into today's high-end routers, routing hubs, and network switches. Some new routers have native multicast packet routing. Multicast-capable routers communicate with neighboring multicast routers and exchange information about group membership and network topology. However, there are many routers that cannot route multicast packets correctly.

A host sends a multicast message out onto a host network where multicast-enabled routers pick it up and forward the message to the appropriate group. Routers keep track of multicast groups dynamically and build distribution "trees" that chart paths from each sender to all receivers. Multicast routers need to be able to execute a multicast routing protocol that defines delivery paths that enable the forwarding of multicast datagrams across an internetwork.

Routers refer to the specific tree that it built for a sender when it receives traffic from the sender for a multicast group. The IP standards bodies have designed several routing protocols that can build the distribution trees for multicast traffic and multicast routers must support one or several of these protocols.

The current routing protocols for multicast include the:

- ✔ **Distance Vector Multicast Routing Protocol (DVMRP)**—An interior-gateway protocol

- ✔ **Multicast Open Shortest Path First (MOSPF)**—An extension to the OSPF link-state unicast routing protocol
- ✔ **Protocol Independent Multicast (PIM)**—This enables networks running any unicast routing protocol to support IP multicast. PIM has two modes, dense and sparse.

Usually multicast protocols can be layered on top of existing multiprotocol backbones with a software upgrade to existing routing devices. Ideally, network routing devices should provide all of these standards: IGMP, DVMRP, MOSPF, and PIM, thereby allowing the widest range of multicast operations and interoperability.

The most common multicast routing protocol is DVMRP, which works by first broadcasting to all reachable multicast routers. When no members are connected to a branch, the branch is pruned from the tree as shown in Figure 4.8 The broadcast from the source can reach a router from several directions. The multicast router selects the shortest of the routes to the group members.

FIGURE 4.8 Pruning a multicast delivery tree

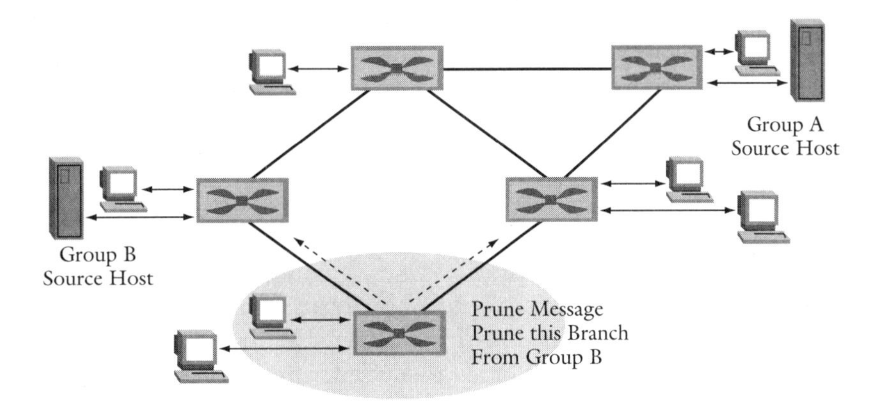

Multicast Routing with IP

The multicast-capable network shown in Figure 4.9 consists of LANs with native multicast support connected by IP multicast-capable routers.

In the Figure 4.9 example, Host A can address the host group by addressing the host group's Class D address. When Host A sends the message out onto its network, the multicast routers pick up the message and forward the multicast transmission to the appropriate subnets.

Each physical network can have several multicast-capable routers. Network protocols select one of them as the designated router for the network. The designated router then communicates with other designated routers in neighboring networks to construct a spanning tree for each multicast source. This procedure is called Reverse Path Multicasting (RPM).) Datagrams from the source host to other group members travel over this spanning tree. A spanning tree is loopless and guarantees the shortest possible route to the receiver.

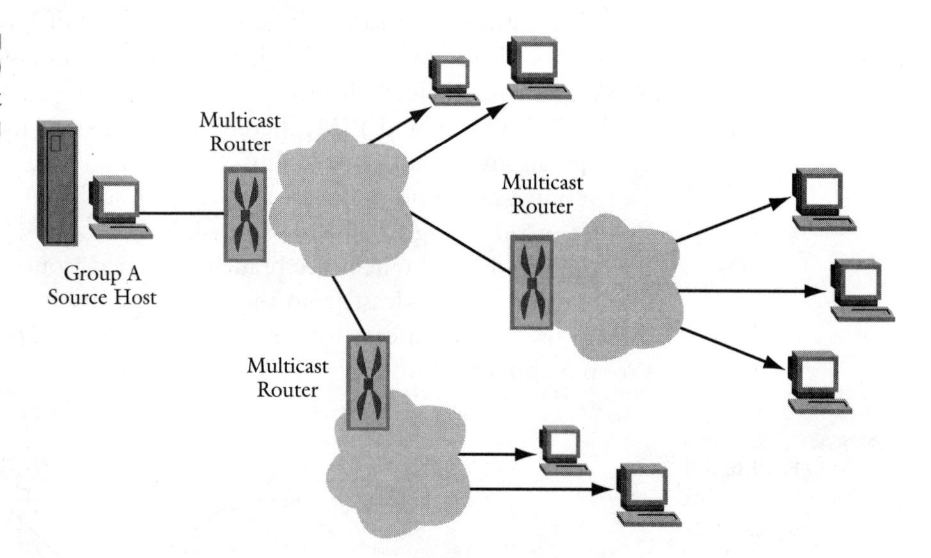

FIGURE 4.9 IP multicast Internet routing

The copying of the message is done by the multicast method of the LANs involved. In this case all LANs support native multicasting and the multicast transmission takes the same bandwidth on host A's network as a single copy, regardless of how many clients are members of the host group on the other side of the Internet. Even if not all LANs have native multicast support, the added cost of transmitting copies will be limited to a single LAN.

When IP multicasting support is added to the IP layer, several potential problems need to be addressed. The major issues include resolving IP multicast addresses to LAN (multicast) addresses, copying and forwarding messages, and registering host group membership.

Older TCP/IP implementations are apt to have no support for sending or receiving multicast transmissions. Such an implementation discards packets addressed to a multicast address as being corrupt or invalid.

For a host to be able to send to a multicast address the IP protocol implementation must support a mapping from the host-group address to the corresponding multicast LAN address. This mapping is analogous to Address Resolution Protocol (ARP) mappings, where host-group addresses are resolved to LAN addresses.

IGMP needs to be implemented so hosts who want to can join host groups and listen to multicast messages. IGMP informs local multicast routers that a host is a member of a particular multicast group. It should be noted that:

- ✔ Hosts on an Internet capable of sending to a multicast address can send packets addressed to a host group address
- ✔ Processes running on hosts capable of receiving multicast packets can join host groups, after which the process receives all transmissions sent to that host group address
- ✔ The routing of IP multicast packets requires special functions absent from unicast IP routers.

Multicast router functions can be located on an IP host. Multicast routers forward multicast transmissions for hosts outside the LAN to the next IP multicast-capable router using a unicast tunneling method. Tunneling is the encapsulation of multicast packets using a unicast IP header between two multicast routers.

IP Multicast-Enabled Multimedia Applications

Multimedia applications are capable of streaming full-motion video, audio, and animated graphical content to the desktop. Other multimedia applications are real-time interactive applications, such as desktop video, collaborative engineering, and shared whiteboards. Multimedia applications and real-time interactive applications such a video conferencing use a great deal of bandwidth. The importance of clear picture and hi-fi quality voice should not be underestimated even in standard everyday videoconferences.

Sometimes there is a need to transmit stored data streams to large numbers of recipients. Examples here could include updates of Web caches or corporate announcements to employees. In this case, using IP multicast-enabled applications would be beneficial and cost-effective.

IP multicast can unite collaborative workgroups and help in the realization of the full potential of these new applications for the workplace.

IP Multicast-Enabled Information Distribution Applications

Information distribution applications benefit from the use of IP multicasting. This category of application can provide real-time news and financial services to the desktop subscribers.

Non-multimedia applications that involve the transfer of large databases of information can benefit from IP multicast in the same way multimedia applications can.

TIP

The IP Multicast Initiative can provide information about IP multicast technologies, products, and services. Visit their Web page at **http://www.ipmulticast.com** or contact them at: Stardust Technologies, Inc., 1901 S. Bascom Ave, #333, Campbell, CA 95008 Tel: 408-879-8080 Fax: 408-879-8081

The IP Multicast Initiative recommends:

- ✔ A phased approach to evaluating and deploying IP multicast including deploying IP multicast on the Intranet before deploying it on the WAN
- ✔ A network analysis and user profile to determine the benefits and costs of implementing IP multicasting for the company
- ✔ Examination of several alternative plans for the deployment of multicast-based applications over a WAN and the use of transitional approaches such as tunneling
- ✔ The use of a test-bed evaluation of both LAN and WAN implementations in-house or at a vendor's site
- ✔ Staff training in IP multicast network administration and diagnosis
- ✔ The choice of IETF standards-based products designed for native IP multicast.

The IP Multicast Initiative Web site has a technical resource center that provides more background and in-depth information.

When deploying IP multicasting; make sure that:

- ✔ The sending and receiving nodes' operating systems and TCP/IP stacks support multicast and the IGMP. The latest versions of UNIX, Windows NT, and Windows 95 accommodate the TCP/IP stack, IP multicasting, and IGMP.
- ✔ Each node's network adapter driver implements multicasting. Newer adapters and network drivers have built-in support for IP multicast.
- ✔ Routers, bridges, and switches in the network support multicasting at the IP layer. Many new routers already support IP multicast; others may require an upgrade. Many network equipment vendors have IP multicast-ready products available.
- ✔ Protocols are chosen carefully; depending on the underlying routing protocol, different protocols may apply.

✔ Applications to be used are multicast-enabled. Standards are being developed and incorporated into applications for enhanced services like reliable and real-time delivery of IP multicasting traffic. Use IP multicast upgrade and enhanced application program interfaces (APIs) available to add IP multicast to applications. IP multicasting APIs include Berkeley sockets multicast API and the Winsock API for Windows applications.

IPv4 versus IPv6: The Multicast Addressing Issues

The IPv6 protocol fully supports IP multicast. As IPv6 becomes implemented in more and more nodes connected to the Internet, multicasting will be integrated. Its use will no longer be limited to local networks and experimental Internet implementations.

TIP

For more information on IPv6, see my book, co-authored with Kitty Niles IPv6 Networks (McGraw-Hill).

IPv6 multicast addresses have the format shown in Figure 4.10. A multicast address can be assigned to a single system, restricted to a specific site, associated with a particular network link, or distributed worldwide. Multicast addresses must not be used as source addresses in IPv6 datagrams or appear in any routing header. A value of FF (11111111) identifies an address as a multicast address. The IPv6 multicast address has three other fields, the `Flgs` field, `Scop` field, and `Group ID` field.

FIGURE 4.10 Multicast address format

8 bits	4 bits	4 bits	112 bits
11111111	Flgs	Scop	Group ID

Flgs:

0	0	0	T

A low-order bit of flags indicates a permanently- or non-permanently-assigned multicast address. Other flag bits are zero. The `Scop` field limits the scope of the multicast group. Possible scope values include node-local, link-local, site-local, organization-local, and global. Non-permanently-assigned multicast addresses are meaningful only within a given scope.[2]

Mapping of IPv6 multicast addresses to Ethernet MAC addresses is similar to IPv4 mapping, but the low-order 32 bits of group addresses are mapped to the MAC address instead of the low-order 23 bits in the IPv4 specification.

What's Next

This chapter discussed multicasting in workgroups, some of its capabilities on hosts and routers, as well as usage and implementation, especially with VoIP. The next chapter introduces the basic concepts of VoIP, and its most used H.323 standard. It also discusses other standards and technologies such as audio codecs, IP over ATM, voice over ATM, the emulation of traditional T1/E1 trunks, IP over SONET and voice over SONET, and IP and voice over frame relay. In addition, it treats Layer 3 switching and Gigabit Ethernet as well as their roles in VoIP.

---

2 For more information about IPv6 addressing, see RFC 1884 by S. Deering and R. Hinden, "IP Version 6 Addressing Architecture," December 1995.

CHAPTER 5

More on ATM Technologies

This chapter provides a more in-depth discussion of ATM technologies and the IP protocols supporting ATM. This topic is very important for an understanding of the features (and limitations) of using IP over ATM, and more specifically, VoIP.

The chapter is based on a contribution made by Phillip Emer, Associate Director of Advanced Technology Development at North Carolina State University. I thank him for his contribution and deep knowledge on the subject of ATMs and their applications.

For more information on this subject, or to contact Phil, use his e-mail address: **phil@ncstate.net**.

Describing ATM Services and Support

The layered approach to describing ATM, as shown in Figure 5.1, stresses the interactions between the many layers and sublayers and the passing of service data units up and down the layered stack. This kind of approach explains how ATM software and hardware components come together to form an ATM node.

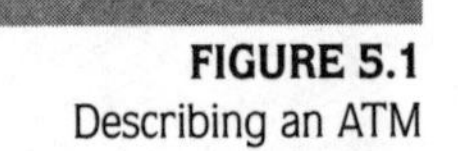

FIGURE 5.1
Describing an ATM

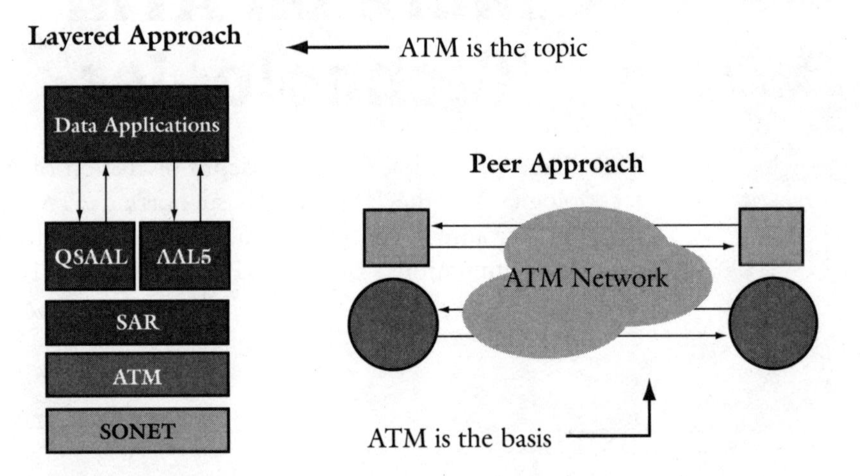

The peer approach to describing ATM addresses the peer-to-peer operations between ATM devices. This approach stresses the functions provided by ATM hardware and software and how these functions interact to support an ATM network. As shown in Figure 5.2, the main characteristics of ATMs relevant to ATM network designers and to the peer approach to ATM in general are:

- ✔ ATMs are connection-oriented and point-to-point
- ✔ They are integrated services—voice, video, data

- ✔ They support QoS
- ✔ They scale to gigabit speeds
- ✔ They use 20-byte addressing.

FIGURE 5.2 Relevant ATM characteristics

Characteristics

- Connection-oriented, point-to-point
- Integrated services—voice, video, data
- Supports quality of service (QoS)
- Scales to gigabit speeds
- Uses 20-byte addressing

A detailed consideration of these characteristics follows.

Connection-Oriented Support

Connection-oriented systems (e.g., POTS, ISDN, X.25, TCP, and ATM), as shown in Figure 5.3, establish connections between a pair of communicating systems. In the phone system, a connection is associated with a physical dedicated link. In ATM, a connection is a virtual link (circuit) which is characterized by a class of service that it must deliver to the users of the circuit (e.g., a constant bit-rate service for uncompressed voice over ATM).

Connection-oriented systems support a set of protocols for transmission of data and another set of protocols for establishing connections (e.g., SS7, Q2931). End systems request that the network establish a connection by sending messages to the network over a predefined control channel (circuit). Once the network has established a data channel (circuit) per the request, data are forwarded over that channel. Connections may be established by a network manager as Permanent Virtual Circuits (PVCs) or may be dynamically established by system software as Switched Virtual Circuits (SVCs).

FIGURE 5.3
A layout of connection-oriented ATM systems

Connection-Oriented

- Systems send connection setup messages over a control channel
- Connection setup may be static (PVC) or dynamic (SVC)
- Data are then transmitted over data channels

Integrated Services Support

As ATM is the outgrowth of broadband ISDN, integrated-services support was a design consideration from the start. It is worth noting though that many technical compromises were necessary to support three very different network-traffic categories. The result of these compromises, listed in Figure 5.4, is that ATM supports the *combination* of voice, video and data optimally—not necessarily the individual services.

FIGURE 5.4
ATMs support integrated services

Integrated Services

- Beneficial to support voice, video and data in a single homogeneous network
- ATM is optimally designed to support the combination of services—not each service independently

Voice Support

Voice is supported in an ATM network via circuit emulation, via circuit switching, or as data. Circuit emulation is a migration path supporting emulation of T1-style WAN circuits, as shown in Figure 5.5. In general many voice circuits are mapped to single ATM PVC. Voice compression, network echo cancellation, and silence-suppression mechanisms may be added as enhancements for bandwidth efficiency. Optimal support of voice would map traditional 64-kbps voice circuits dynamically to ATM circuits (SVCs).

FIGURE 5.5
ATM's voice support via circuit emulation

Voice

- **Circuit emulation:** Many voice circuits mapped to a single ATM circuit (PVC)
  - Uncompressed: 64 kbps
  - Compressed: 8–32 kbps
    - ADPCM, CELP, LD-CELP
    - ITU—G.721, G.728, G.729
- **Circuit switched:** Voice circuits map directly to ATM circuits
- **As data:** Voice over IP

Voice is also supported on IP platforms as data. Some of the same coding, compression, and echo-cancellation techniques as in the circuit emulation case are applied here.

Video Support

Video is supported, as shown in Figure 5.6, as real-time, non-real-time, and as data. Real-time video applications include video conferencing and live video broadcasts. Support of real-time video levies strict delay requirements on the ATM network (actually strict delay variation—or jitters—requirements).

Non-real-time video applications such as video-on-demand use application-level techniques—namely buffering—to account for inconsistent delays encountered during transit through the ATM network. Thus the ATM network need not be responsible for ensuring low delay variation requirements (at least not as strictly as in the real-time case.

FIGURE 5.6 ATMs support video as real-time and non-real-time

Video

- **Real-time:** Time dependent as in conferencing or live television
- **Non-real-time:** Not time-dependent as in video or on-demand
- **As data:** ITU H.323, IP-based video conferencing

Nonetheless, video over ATM is a bit more complicated to discuss than voice, since video terminals are actually audio/visual terminals, that is, voice and video are related and must be kept synchronized. Figure 5.7 illustrates a simple representation of a broadband (ATM) video terminal (H.310 in ITU jargon).

FIGURE 5.7 A simple representation of a broadband (ATM) video terminal

Video

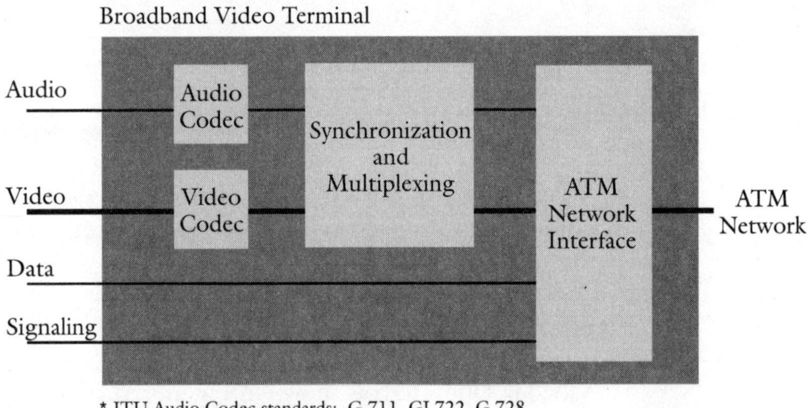

* ITU Audio Codec standards: G.711, GJ.722, G.728
ITU Video Codec standards: H.261, H.262

Data Support

Data can be supported, as shown in Figure 5.8, at various levels: LAN, IP, Multi-protocol, and Native.

FIGURE 5.8
Data support on ATMs

Data

- **LAN:** Support NOS applications
- **IP:** Support IP applications
- **Multi-protocol:** Support all applications
- **Native:** Rewrite applications to take advantage of QoS and account for signaling

The idea of supporting data over ATM is that a migration path is required since data applications are closely tied to platforms, operating systems (OSs), application programming interfaces (APIs), and protocol stacks (e.g., IP, IPX, Appletalk), as shown in Figure 5.9:

1. First, software shims are added around low-layer APIs (close to device drivers)—the effect is to hide ATM from existing applications and protocols (users).
2. Then, low-layer and higher-layer APIs are enhanced to incorporate software shim function and to expose ATM features to existing stacks.
3. Finally, a new stack and new data applications emerge—all of which are ATM-aware.

Quality of Service Support

As Figure 5.10 outlines, ATM supports QoS, enabling end-to-end integrated service support, guaranteeing performance to connections and service classes (queues), including:

- ✔ Constant bit rate (CBR)
- ✔ Variable bit rate (VBR)—real-time and non-real-time
- ✔ Unspecified bit rate (UBR)
- ✔ Available bit rate (ABR).

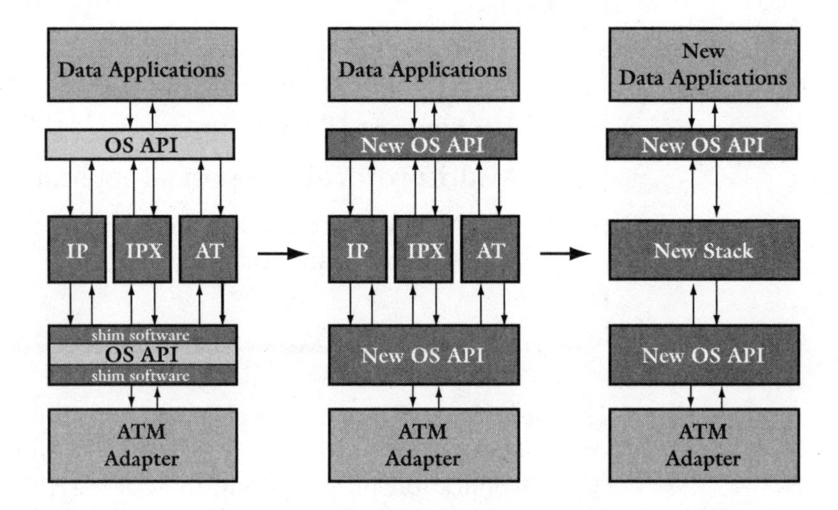

FIGURE 5.9 ATM's data support requires a migration path since data applications are closely tied to platforms

FIGURE 5.10 ATM's QoS support

Quality of Service

- Enables end-to-end integrated service support
- Guarantees performance to connections
- Service classes (queues) include
  - Constant bit rate (CBR)
  - Variable bit rate (VBR)—real-time and non-real-time
  - Unspecified bit rate (UBR)
  - Available bit rate (ABR)

ATM technology is also scalable to gigabit speeds (as Figures 5.11 and 5.12 outline). ATMs also use 20-bit addressing, as shown in Figure 5.13:

- ✔ The Network Service Access Point (NSAP) prefix identifies an ATM network
- ✔ The End System Identifier (ESI) identifies a device on an ATM network
- ✔ Selector (Sel) allows a device with a single ATM interface to have multiple ATM addresses.

FIGURE 5.11
ATM support scaled to gigabit speeds

Scales to Gigabit Speeds

- Ethernet, FDDI, and token ring physical and media access layers limit attainable network speed, size, and distance
- No new development for TR and FDDI
- Gigabit ethernet uses a different physical layer and is distance-bound by CSMA/CD

FIGURE 5.12
ATM's interface speeds

Interface Speeds

- SONET physical layer
  - 155 Mbps (OC3)—backbone, server
  - 622 Mbps (OC12)—backbone
- Copper (UTP) physical layer
  - 10 Mbps—desktop*
  - 25 Mbps—desktop
  - 51 Mbps (OC1)—residential
  - 155 Mbps (OC3)—workstation, server

*ATM over Ethernet—aka, cells in frames (CIF)

ATM network addressing is similar to Novell Netware network addressing in that the end system has an address with two distinct pieces—a network piece and a local piece, as shown in Figure 5.14. It turns out this is a nice feature, especially when analyzing and troubleshooting a network. Since ATM addresses are made up of a network part and a local part, a misbehaving client can be tracked down easily based on its ATM address.

ATM Address Format

- Network Service Access Point (NSAP) prefix identifies an ATM network
- End System Identifier (ESI) identifies a device on an ATM network
- Selector (Sel) allows a device with a single ATM interface to have multiple ATM addresses

FIGURE 5.13 ATM address format

ATM Address Format

- ATM adapters shipped with a fixed ESI
- ATM adapter and switch exchange management information at startup—ILMI (Interim Local Management Information)
- ATM adapter "gets" NSAP prefix from switch and appends ESI/Sel to yield its ATM address (IPX-like)

FIGURE 5.14 ATM network addressing is similar to Novell Netware network addressing

Try to contrast this to a misbehaving IP network host. The IP addressing scheme is totally distinct from the LAN addressing scheme—so knowing an IP address is not enough to isolate a misbehaving client in this case (the local address of the client must be known).

Figure 5.15 shows that an ATM address can be represented in a naming format as an IP address can. Note that ATM device names will use the same domain name services (DNS) that IP hosts use—with an extension that maps 20-byte local addresses to a name.

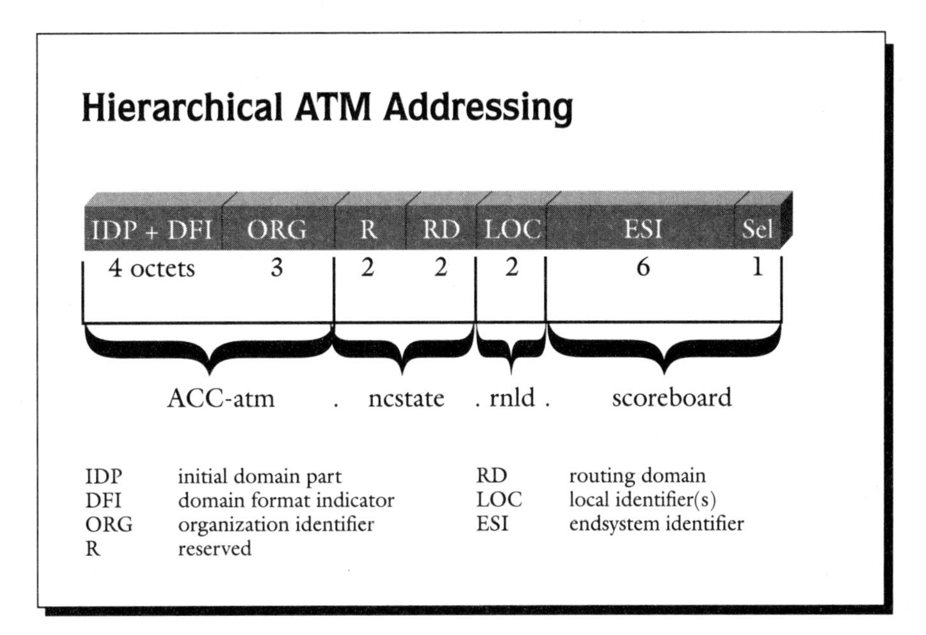

FIGURE 5.15 An example of a hierarchical addressing structure

As is the case for IP addresses, there is a process in place for network managers to acquire unique network name and address space. In the case of ATM there are several registration authorities from which to acquire ATM code points (in the U.S.)—the National Institute of Standards and Technology (NIST) and the American National Standards Institute (ANSI). Basically, ANSI or NIST assigns a three-octet organization identifier (ORG) code point. This ORG field follows four octets of identification bits, which identify the country and registration authority (e.g., ANSI). Once an organization obtains an ORG value, the seven most significant octets of the ATM address uniquely identify the ATM network of that organization. The owning organization is then responsible for the encoding of the remaining six octets in the ATM-network part of the ATM address. Recall that the least significant seven octets of the ATM address are locally (client) significant and associated with the ATM client device.

Summary

In summary, as Figure 5.16 indicates,

- ✔ ATM is flexible—it provides integrated services
- ✔ ATM is scalable—it enables SONET transmission and 20-byte addressing
- ✔ ATM is available.

FIGURE 5.16
Summary of benefits of ATM technologies

Summary

- ATM is flexible—integrated services
- ATM is scalable—SONET transmission, 20-byte addressing
- ATM is available

ATM Networking

When it comes to deploying ATM networks, there are several models which can be adopted. Figure 5.17 outlines the main models: end-to-end, desktop, campus, WAN, and carrier.

FIGURE 5.17
ATM networking: a list of the main models

ATM Network Background

- The end-to-end ATM model
- ATM desktop model
- ATM campus model
- ATM WAN model
- ATM carrier model

The End-to-End ATM Model

This chapter has already discussed one of the relevant characteristics of ATM: it supports integrated services. ATM can support voice, video, and data in the campus, in the WAN, in the carrier network, and in residential access networks, as Figures 5.18 and 5.19 show.

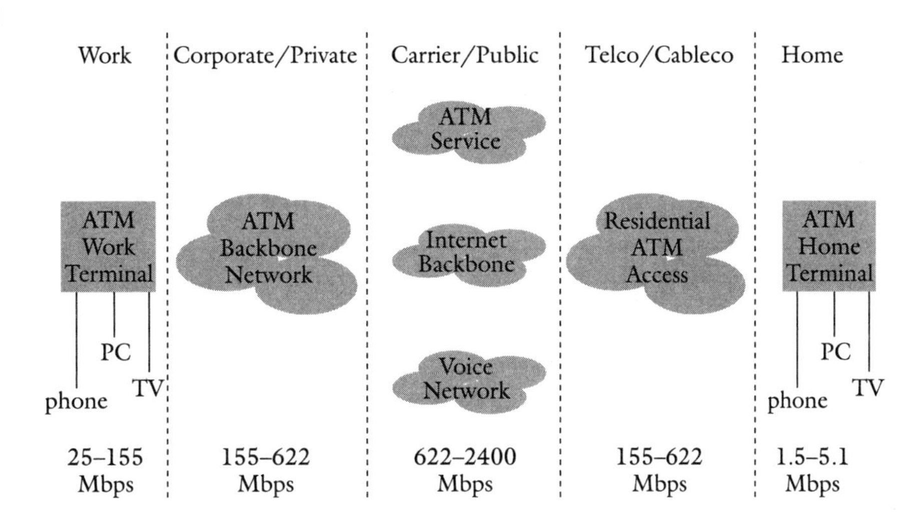

FIGURE 5.18 The end-to-end ATM model

FIGURE 5.19 An outline of the applications of end-to-end ATM model

End-to-End ATM Model

- ATM is at the core of public and private networks
- Voice, video, and data flow over ATM, *integrated*, all the way to the user
- ATM circuits terminate in PCs, set-top boxes, "home terminal" devices...end-points

At the moment ATM is implemented in campus, WAN, and Internet backbones, as depicted in Figure 5.20. Figure 5.21 provides the status of end-to-end ATM at glance. There have been some ATM-based trials in residential-access environments (cable modems and ADSL)—commercial services are likely to appear over the next year mainly for Internet access.

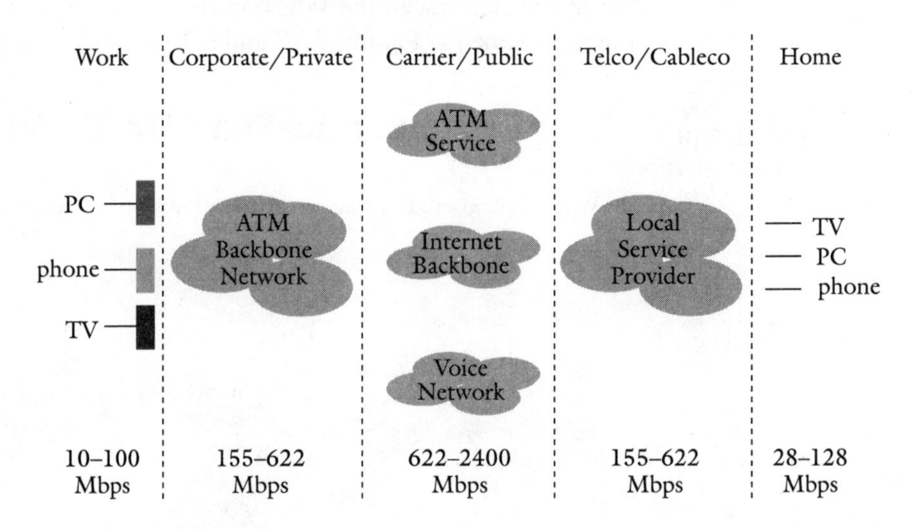

FIGURE 5.20 At the moment, ATM is being implemented in campus, WAN, and Internet backbones

FIGURE 5.21 Status of ATM end-to-end deployment to date

End-to-End ATM – Status

- ATM is at the core of public and private networks
- Voice, video, and data flow over ATM, integrated, *in the core of the network*
- ATM circuits terminate in ATM *adaptation devices, e.g., routers and codecs*

The ATM Desktop Model

In the ATM desktop model, as depicted in Figure 5.22, workgroup switches are replacing Ethernet hubs. Category 3/5 cabling is being used for the LAN emulation in providing migration path. Also, old applications are being rewritten to take advantage of QoS, thus enabling high-quality video and voice to join data. Figure 5.23 shows a status of ATM desktop model implementations.

FIGURE 5.22
The ATM desktop model

The Desktop Model

- Workgroup switches replace Ethernet hubs
- Category 3/5 cabling
- LAN emulation provides a migration path
- Old applications rewritten to take advantage of QoS
- High-quality video and voice join data

FIGURE 5.23
Status of implementations of the ATM desktop model

The Desktop Model – Status

- Ethernet switches replace Ethernet hubs
- Category 3/5 cabling
- LAN emulation is a *VLAN* solution
- *No* applications rewritten to take advantage of QoS
- Video and voice *are* data

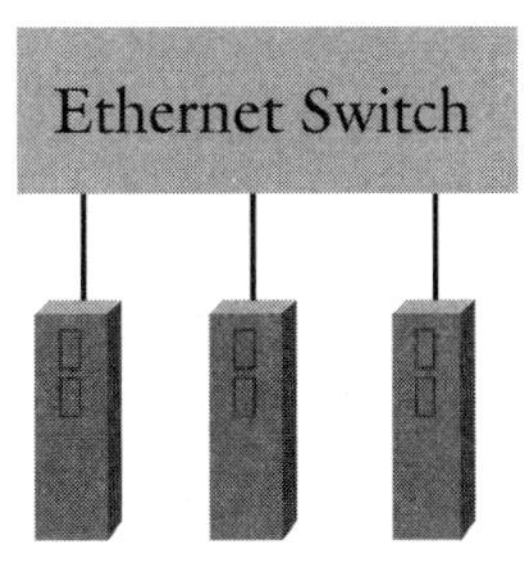

The Campus Backbone Model

Campus backbone models are using ATM switching to replace FDDI in the backbone. Video and voice networks are migrating to ATM as the data network is flattened. We also see that VLANs obviate the need for routing, as switched virtual circuits dominate. Figure 5.24 provides an outline of these applications and Figure 5.25 shows the status of their implementation.

FIGURE 5.24
The ATM campus backbone model

Campus Backbone Model

- ATM switching replaces FDDI in the backbone
- Video and voice networks migrate to ATM
- Data network is flattened
- VLANs obviate the need for routing
- Switched virtual circuits dominate

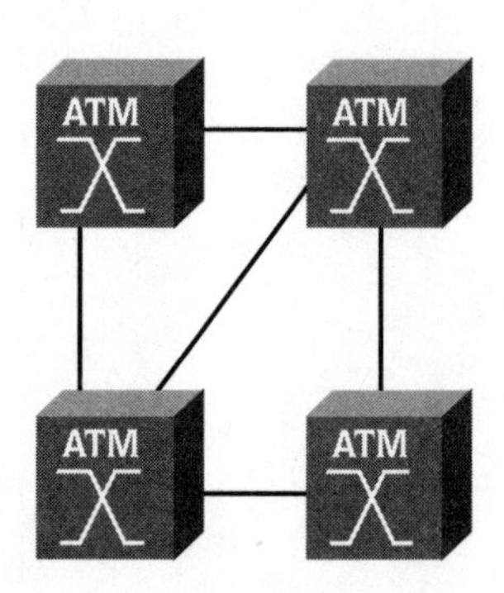

FIGURE 5.25
Status of implementation of the ATM campus backbone model

Campus Backbone – Status

- ATM switching is the only *alternative* to FDDI
- Video and voice run *over IP as data*
- Data network is *the same*
- VLANs provide broadcast control in *flat networks*
- Switched virtual circuits dominate (*performance*)

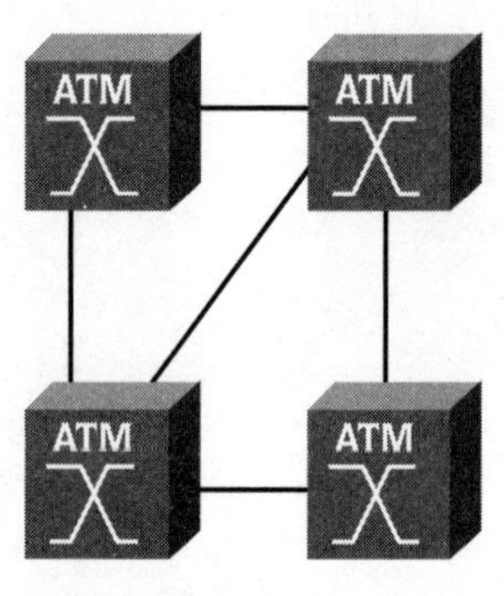

The WAN Model

In the WAN model, as shown in Figure 5.26, the idea is to replace TDM-type multiplexers with ATM edge switches. These edge switches support adaptation of LAN data traffic and PBX voice traffic to ATM.

Also, the idea here is to use the same WAN links to carry voice and data—ATM will allow for running these links at much higher utilization due to the QoS and backbone WAN (BW) management capabilities inherent in ATM. The ATM adaptation for voice in this case comes in two flavors: uncompressed constant bit rate, and compressed (e.g., ADPCM, CELP) variable bit rate.

FIGURE 5.26
The ATM WAN model

The WAN Model

- ATM edge switches replace TDM multiplexers
- ATM BW management allows for voice and data to share links
- Result is significant decrease in monthly WAN link costs

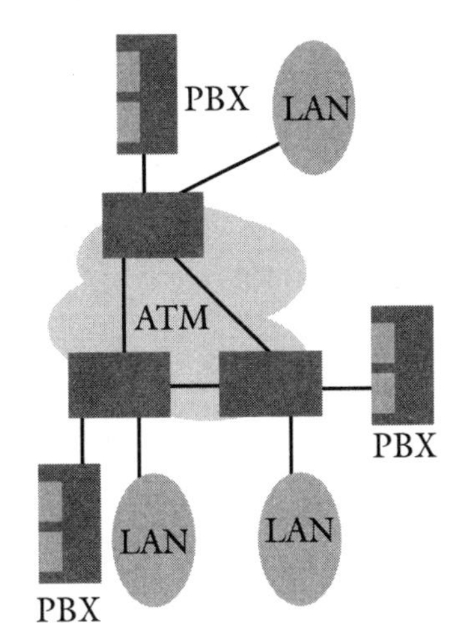

WAN network managers are accustomed to the "over-subscription" method of BW management. Moving to ATM in this arena requires a departure from this inefficient method of BW management, as depicted in Figure 5.27. Like any paradigm shift it will take time.

FIGURE 5.27
Status of ATM WAN deployment

The WAN Model – Status

- ATM edge switches replace TDM multiplexers
- ATM BW management *is a new paradigm for network managers*
- Result is *cautious adoption*

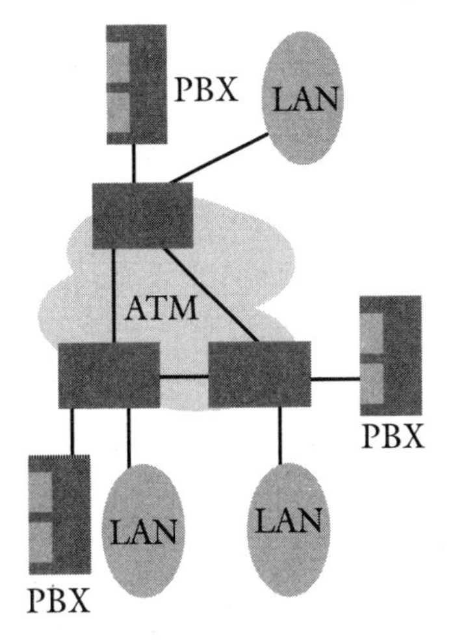

ATM Carrier Model

Voice in the carrier model means switching voice circuits as ATM circuits, unlike the WAN ATM voice scenario, as depicted in Figure 5.27, which may switch many voice circuits over a single ATM circuit.

FIGURE 5.28
ATM carrier model

ATM Carrier Model

- Switches voice circuits
- Serves as new POTS infrastructure
- ATM carrier networks serve as transport for compressed video
  - Video on demand
  - Video conferencing
  - Distance education

Figure 5.29 shows the status of implementations over ATM carrier models. Time Warner implemented an ATM-based, interactive, video-on-demand service, as a trial, in Orlando, Florida. In that trial ATM circuits terminated in a set-top box—ATM transported MPEG2 video to the set-top. Time Warner found that it is better to use ATM switching in the carrier network and use MPEG2 as the transport for video. Another, perhaps more important, finding was that users are not willing to pay enough money for interactive television to justify large-scale implementation of such systems. Instead, carriers, telcos, and cable companies are focusing on offering Internet-based interactive services to the home. Much of the work in this area employs ATM switching and in some cases ATM connections to the home. The Remote Broadband working group in the ATM Forum writes specifications for ATM-based residential networking (e.g., ATM-based cable modems and ADSL).

In summary, Figure 5.30 indicates:

- ✔ Ethernet dominates desktop access
- ✔ IP dominates as data transport
- ✔ IP transports voice and video as data
- ✔ ATM switches IP
- ✔ Adaptation devices at the edge of ATM networks "adapt" voice, video, and data individually to flow optimally across ATM networks.

FIGURE 5.29 ATM carrier model implementation status

ATM Carrier Model – Status

- No voice switching *yet*
- Only a few vendors offer central office switches
- ATM carrier networks *switch* compressed video
  - Video on demand (IP)
  - Video conferencing (IP)
  - Distance education (IP)

FIGURE 5.30 Summary of ATM carrier model deployment

Summary

- Ethernet dominates desktop access
- IP dominates as data transport
- IP transports voice and video as data
- ATM switches IP
- Adaptation devices at the edge of ATM networks "adapt" voice, video, and data individually for optimal flow across ATM networks

IP Over ATM Requirements

As IP is the dominant network layer data protocol, ATM network designers are clearly interested in the requirements associated with using ATM as a transport for IP data. More specifically, they are interested in how IP uses the services of ATM (or perhaps more correctly—how to hide the characteristics of ATM from IP).

In general, as shown in Figure 5.31, emerging IP over ATM (IPoATM) protocols must address:

✔ Encapsulation (and multiplexing)
✔ Emulating point-to-point links (no more shared media)

- ✔ Establishing connections (since IP is connectionless)
- ✔ Integration with legacy networking
- ✔ Internetworking (routing issues).

FIGURE 5.31
IP over ATM requirements

Requirements

- Encapsulation of frames/packets
- Emulate point-to-point links
- Connection maintenance
  - Signaling
  - ARP services
  - Multicast services
- Integration with legacy networking
- Inter-network communications (routing)

Encapsulation issues include how packets are mapped to cells and how the IPoATM protocol recognizes a stream of ATM-encapsulated packets. Emulating a point-to-point link means treating a virtual circuit as a dedicated link. Establishing connections involves resolving IP characteristics into ATM connections in support of unicast and multicast operation. Integration with legacy networking involves bridging and gateway strategies. Finally, internetworking in the IPoATM context involves defining "a network" and providing a mechanism for communications between these networks (i.e., routing).

Chapter 3 discussed how IETF and the ATM Forum are contribution-driven groups dedicated to the promotion of openness and interoperability in the IP and ATM marketplaces, respectively. In Figures 5.32 and 5.33 it is easy to note that both the IETF and the ATM Forum create interoperability specifications, although both groups argue that they do not write standards.

Figure 5.34 provides a layered model to illustrate the similarities and differences between the pot luck of IPoATM protocols.

FIGURE 5.32
IP over ATM specifications

Specifications

- IETF
  - RFC 1577—Classical IP (CIP)
  - RFC 1483—Multiprotocol encapsulation
  - Next-Hop Resolution Protocol (NHRP)
  - Multiprotocol Label Switching (MPLS)
- ATM Forum
  - LAN Emulation (LANE)
  - Multiprotocol Over ATM (MPOA)

FIGURE 5.33
The data protocols

The Data Protocols

- RFC 1483—Multiprotocol encapsulation
- RFC 1577—Classical IP (CIP)
- Next-Hop Resolution Protocol (NHRP)
- ATM Forume LAN Emulation (LANE)
- ATM Forum Multiprotocol Over ATM (MPOA)
- IP Switching and Multiprotocol Label Switching (MPLS)

ATM Network Services

An ATM backbone network must at minimum support a signaling service, as shown in Figure 5.35. Such a signaling service allows clients (users) to establish connections across the connection-oriented ATM "cloud". At present two signaling services are available for campus ATM backbones—the ATM Forum-defined UNI signaling and the IETF (Ipsilon)-defined general switch management protocol. All switches support UNI signaling and many (not all) support GSMP signaling as well.

FIGURE 5.34
The ATM data model

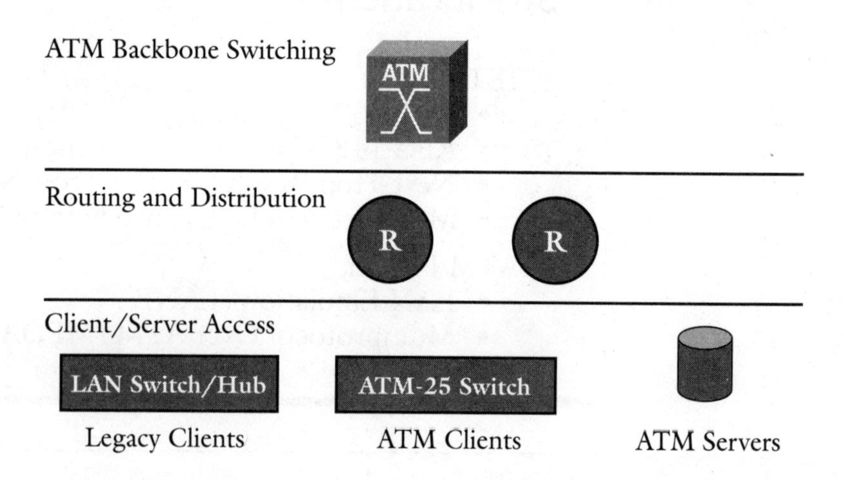

The ATM network may also support address resolution services, multicast services, routing services, etc.—in general, services necessary to support legacy LAN applications (e.g., IP applications like ftp and Telnet).

FIGURE 5.35
ATM network services

Network Services

- Signaling
  - ATM Forum UNI 3.x/4.0
  - General switch management protocol (IETF RFC 1987)
- Others necessary to support legacy protocols
  - Address registration
  - Multicast

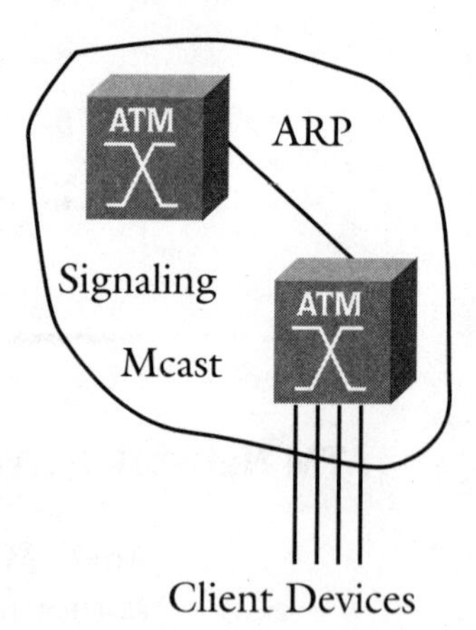

In an ATM environment virtual circuits emulate links, as shown in Figure 5.36. Virtual circuits then act as virtual links, which implies virtual ports and logical networks. At the IP layer the logical view of the network may be quite different from the physical topology.

FIGURE 5.36
ATM routing functions

Routing Functions

- Provide routing services to ATM network
- Provide connectivity to legacy devices and networks

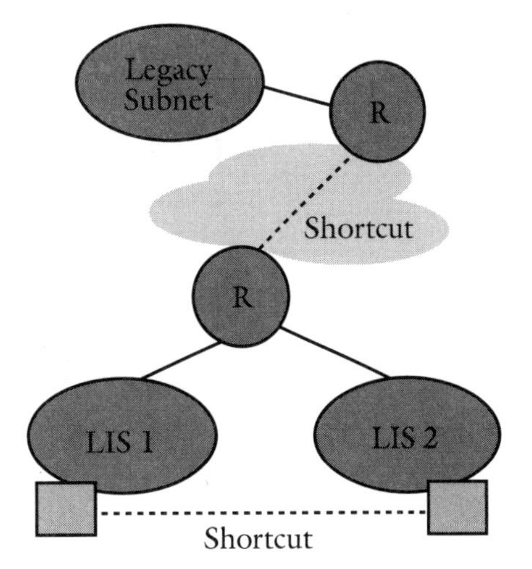

As we deploy fast LAN switching at the access point of the network and fast ATM switching at the core of the network, the software-intensive IP functions performed by routers are clearly the bottleneck. Recognizing that IP routing function is required, we try to avoid forwarding packets through routers while benefiting from routing services.

ATM Client/Server Accessibility

Figure 5.37 describes ATM's client/server accessibility. Note that this is not an exhaustive list of client devices. There are other ATM-attached devices that perform adaptation, such as the cells in frames attachment device (CIF-AD), which works in concert with Ethernet clients to provide an ATM over Ethernet function. An ATM concentrator is an example of an ATM-attached device that performs no adaptation (ATM WG switches are used in place of these devices today).

Other interesting access devices include cable modems, xDSL modems, and inverse multiplexing access devices. If we look at the list of interoperability data protocols in Figure 5.38, and pin down the RFC 1483, Multiprotocol Encapsulation, notice that routers are the only ATM-attached devices in this model, as illustrated in Figure 5.39.

FIGURE 5.37 ATM's client/server accessibility

Client/Server Access

- ATM-attached LAN devices and endstations adapt legacy application behavior to ATM network
- ATM workgroup switches are ATM-only
- Legacy attached devices rely on a higher-layer device to do adaptation

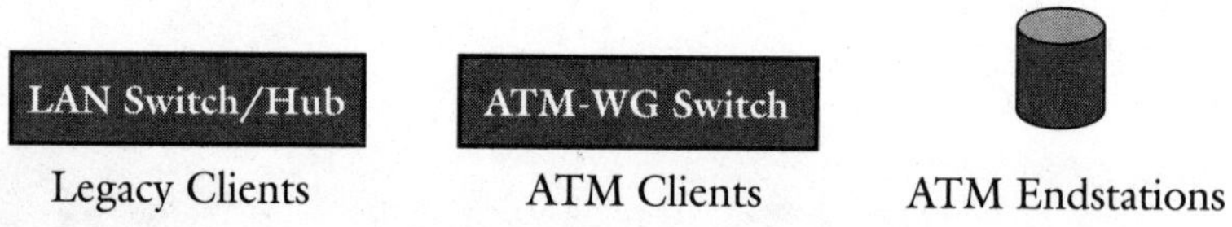

LAN Switch/Hub	ATM-WG Switch	
Legacy Clients	ATM Clients	ATM Endstations

FIGURE 5.38 RFC 1483, Multiprotocol encapsulation

The Data Protocols

- **RFC 1483 – Multiprotocol Encapsulation**
- RFC 1577 – Classical IP (CIP)
- Next-Hop Resolution Protocol (NHRP)
- ATM Forum LAN Emulation (LANE)
- ATM Forum Multiprotocol Over ATM (MPOA)
- IP Switching and Multiprotocol Label Switching (MPLS)

Also notice in Figure 5.40 that RFC 1483 defines an encapsulation method for supporting multiprotocol encapsulation over ATM AAL5. RFC 1483 provides two methods for supporting this encapsulation. LLC/SNAP encapsulation uses IEEE 802.2 SAP headers for multiplexing, protocols over a single ATM circuit. The second method proposes distinct virtual circuits for each protocol.

If we look at RFC 1577 (see Figure 5.41), Classical IP (CIP), notice that the ATM backbone supports several services in addition to signaling, as described in Figure 5.42. CIP relies on an one-armed router for routing services. CIP also requires a router for communication with legacy LAN devices, which makes sense since CIP supports only IP. Thus, LAN clients, such as servers and workstations, communicate with CIP clients through a router. Also keep in mind that CIP is a mechanism supporting ***unicast*** IP over ATM which requires one-armed router function for communication between CIP subnets.

FIGURE 5.39
Diagram of RFC 1483

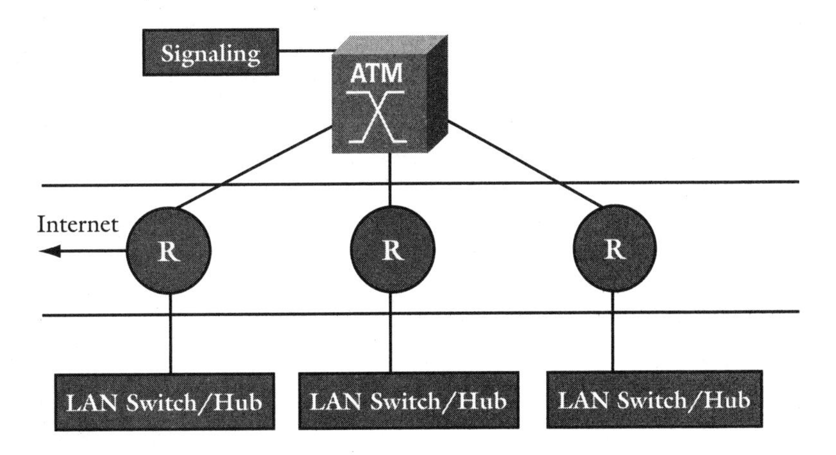

FIGURE 5.40
RFC 1483 specification

RFC 1483

- ATM services—UNI 3.0/3.1 (PVC)
- Routers are ATM-attached
- Defines encapsulation method for muxing protocols over an ATM link
- Makes an ATM circuit look like a point-to-point link (to the router)
- **RFC 1483 is an encapsulation method, *not* a protocol**

FIGURE 5.41
RFC 1577, Classical IP

The Data Protocols

- RFC 1483 – Multiprotocol Encapsulation
- **RFC 1577 – Classical IP (CIP)**
- Next-Hop Resolution Protocol (NHRP)
- ATM Forum LAN Emulation (LANE)
- ATM Forum Multiprotocol Over ATM (MPOA)
- IP Switching and Multiprotocol Label Switching (MPLS)

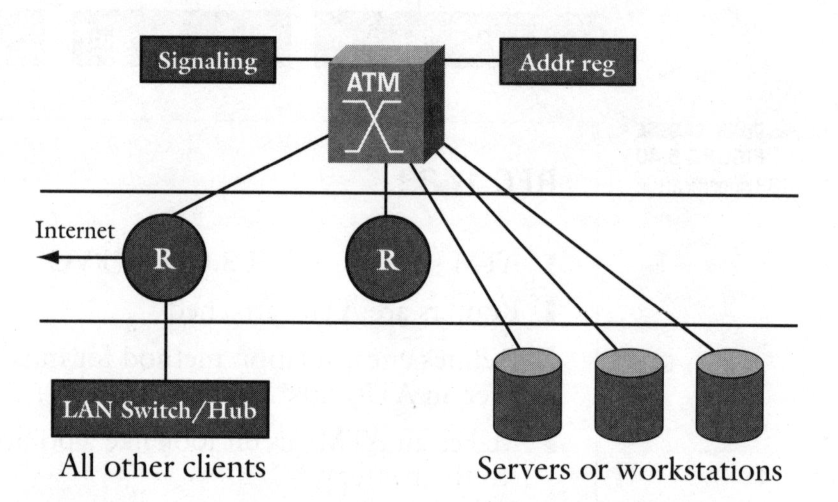

FIGURE 5.42
A diagram of a classical IP (CIP)

ATM Classical IP Services

Classical IP services provide a means of mapping connectionless protocols (the IP protocol), into a connection-oriented backbone environment. To this end CIP provides an ATM service for resolving IP addresses into ATM addresses—thus allowing CIP clients to establish a connection to a destination IP host by using the ATM network's signaling service.

For IP multicast support over ATM, RFC 2022 provides a mechanism for mapping IP multicast addresses to ATM multipoint circuits, as shown in Figure 5.43.

FIGURE 5.43
An overview of ATM CIP services

ATM CIP Services

- ATM Forum UNI 3.0/3.1 signaling
- RFC 1577 ARP services
    - map ip@<-->atm@
- RFC 2022—Adds multicast address resolution (MARS)

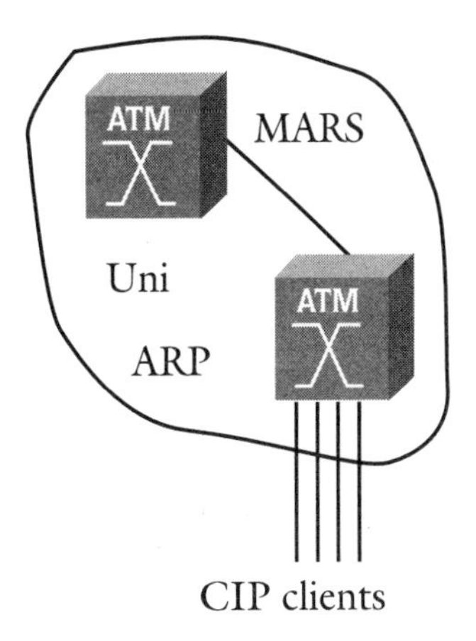

An one-armed router provides routing services for the ATM backbone, as described in Figure 5.44. An one-armed router can also provide routing services between CIP logical IP subnets (LISs) by becoming a member of each LIS. Note that an LIS is defined by a set of hosts registered with the same ARP server.

A one-armed router is so-named because it attaches to the ATM backbone via a single ATM physical link. The router then associates logical ports with the lone ATM interface by joining logical IP subnets. There is then a virtual link associated with each logical port.

FIGURE 5.44
Classical IP routing services

CIP Routing Services

- All legacy devices behind router
- Need one-armed router to talk between CIP logical IP subnets (LISs)

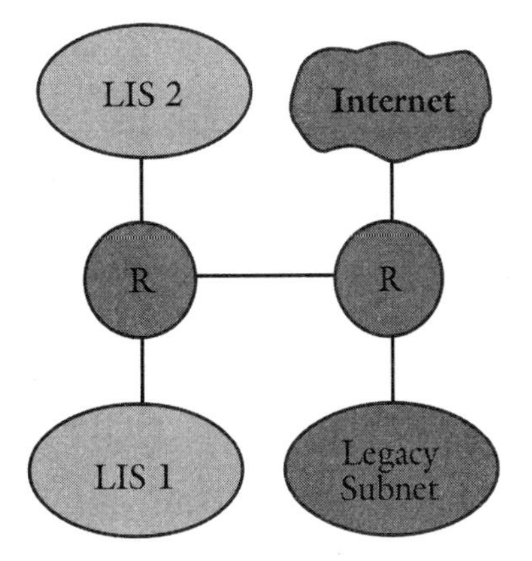

Figure 5.45 shows how legacy IP subnets and hosts, communicate with CIP subnets through an ATM-attached router by emulating a point-to-point link using RFC 1483 encapsulation.

FIGURE 5.45
Classical IP clients

The Data Protocols

- RFC 1483 – Multiprotocol Encapsulation
- RFC 1577 – Classical IP (CIP)
- **Next-Hop Resolution Protocol (NHRP)**
- ATM Forum LAN Emulation (LANE)
- ATM Forum Multiprotocol Over ATM (MPOA)
- IP Switching and Multiprotocol Label Switching (MPLS)

Remember that the classification of RFC 1577 given here applies to RFC 1577 alone. By adding functions specified in new RFCs and Internet drafts into RFC 1577 networks, one can build a more complete system—though other protocols, specially MPOA, integrate all of these functions into a single specification.

The Next-Hop Resolution Protocol (NHRP)

The ATM backbone supports several services in addition to signaling. NHRP (Figure 5.46) adds cut-through routing as an ATM service. LAN clients communicate with NHRP clients through a router, as diagrammed in Figure 5.47.

NHRP enhances CIP function by adding "shortcuts" as outlined in Figure 5.48. That is, the NHRP routing function is provided as a function of the ATM backbone. This routing function allows NHRP clients who are members of different LISs to establish a "shortcut" path through the ATM cloud—recall that with CIP a one-armed router must forward packets between LISs.

FIGURE 5.46 ATM backbone supports the Next Hop Resolution Protocol

CIP Clients

- High-end ATM-attached systems
- Routers – for communication between CIP networks and with legacy hosts
- **CIP is a protocol that allows for high-BW, IP communications between ATM-attached hosts**

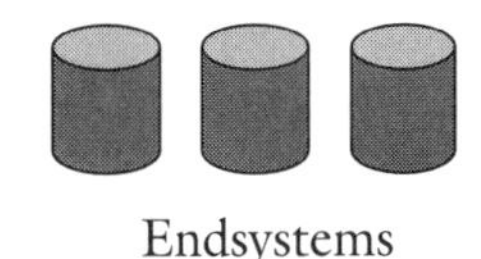

FIGURE 5.47 NHRP adds cut-through routing as an ATM service

Next-Hop Resolution Protocol

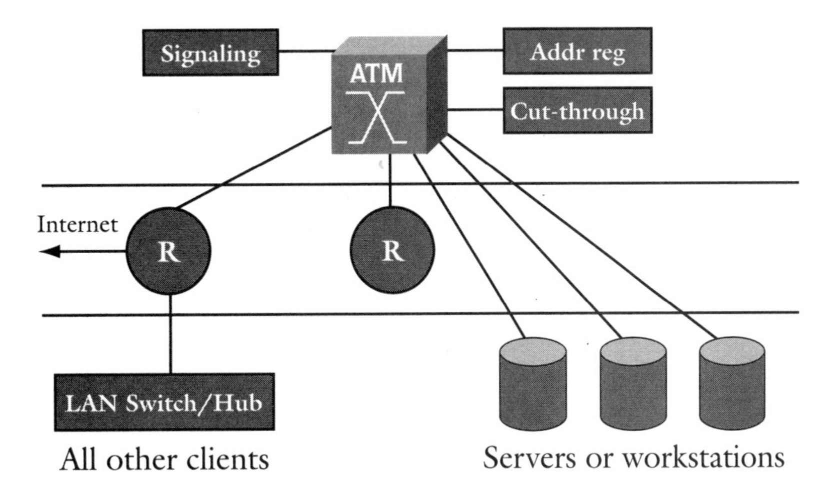

NHRP can also be deployed in an ATM-based IP backbone, as shown in Figure 5.49. In this situation, NHRP provides a shortcut path from one edge of the IP cloud to the other—avoiding interior routing hops. NHRP requires the same address registration function CIP does, as shown in Figure 5.50, and adds distribution of address resolution across LIS boundaries.

FIGURE 5.48
ATM's NHRP services

ATM NHRP Services

- ATM Forum UNI 3.0/3.1 signaling
- NHRP ARP services
  - map ip@<-->atm@
  - Provides "short-cut" routing
- RFC 2022—Adds multicast address resolution (MARS)

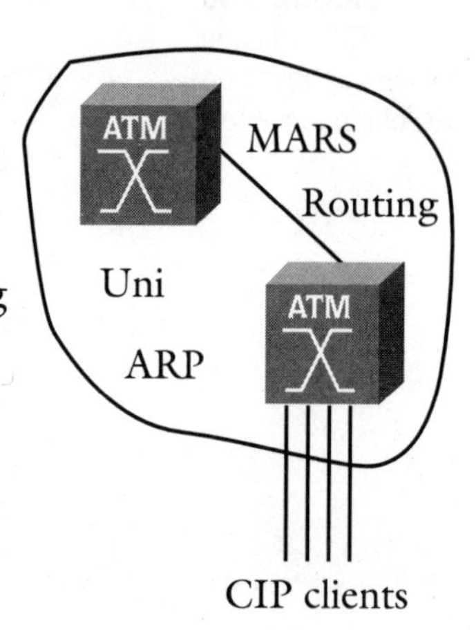

FIGURE 5.49
NHRP routing services

Routing Services

- All legacy devices behind router
- NHRP provides "short-cut" routing between NHRP clients or between entry and exit routers

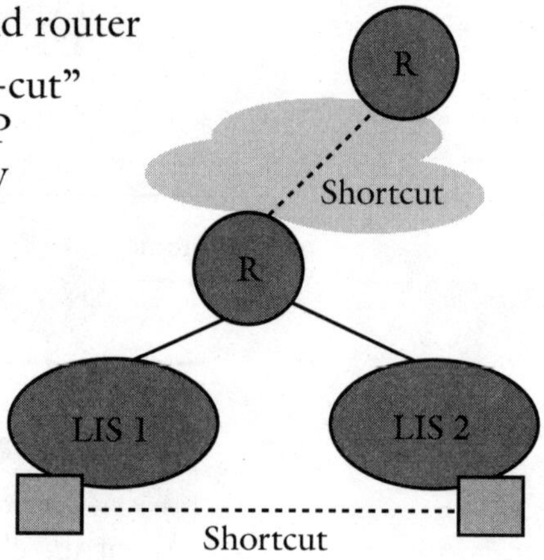

ATM Forum LAN Emulation

Another service in addition to signaling which the ATM backbone supports is ATM Forum LAN emulation (LANE) (Figure 5.51). LANE relies on a one-armed router for routing services. It employs a router for communication with the Internet, as described in Figure 5.52. LANE clients can be LAN switches and ATM-attached stations.

FIGURE 5.50
NHRP clients

NHRP Clients

- High-end ATM-attached systems
- Routers for communication with legacy hosts
- **NHRP is CIP enhanced with support for shortcut routing**

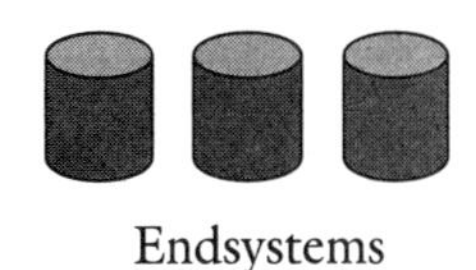

Endsystems

FIGURE 5.51
LAN emulation (LANE)

The Data Protocols

- RFC 1483 – Multiprotocol Encapsulation
- RFC 1577 – Classical IP (CIP)
- Next-Hop Resolution Protocol (NHRP)
- **ATM Forum LAN Emulation (LANE)**
- ATM Forum Multiprotocol Over ATM (MPOA)
- IP Switching and Multiprotocol Label Switching (MPLS)

Figure 5.53 shows how ATM LANE services provide the means of mapping a connectionless, multicast-capable LAN into a connection-oriented backbone environment. LANE provides an ATM service for resolving LAN MAC addresses into ATM addresses—thus allowing LANE clients to establish connections to destination LAN stations by using the ATM network's signaling service. LANE also provides a mechanism for mapping LAN multicast (address-based) into ATM multicast (tree-based)—the broadcast and unknown server (BUS).

FIGURE 5.52 Diagram of a typical LANE

LAN Emulation

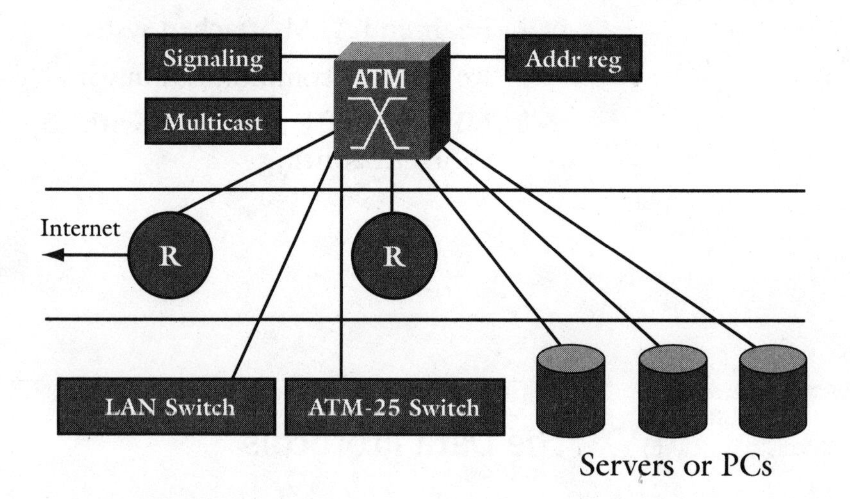

FIGURE 5.53 ATM LANE services

LANE Services

- AF-UNI 3.x/4.0
- LANE Server (LES)—
  mac @<-->atm@
- Broadcast and Unknown Server (BUS)—Mac multicast and default data path

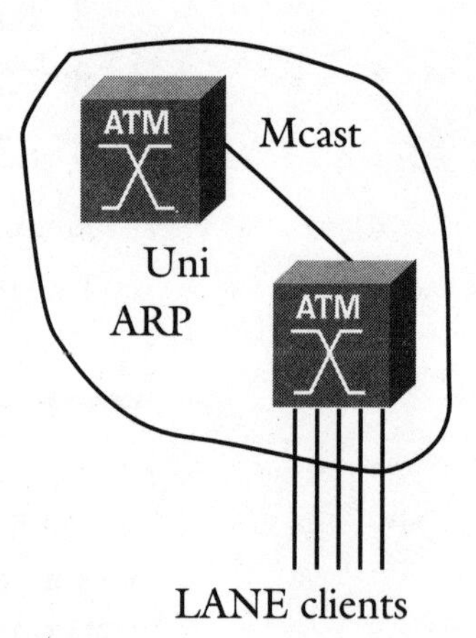

Figure 5.54 shows the LANE routing services' one-armed router that provides routing services for the ATM backbone. An one-armed router can provide routing services between LANE emulated LANs (ELANs) by becoming a member of each ELAN. Note that an ELAN is defined as the set of hosts which register their MAC and ATM address pairs with the same LANE server.

FIGURE 5.54
LANE routing services

Routing Services

- Router to the Internet
- Need one-armed router to talk between LANE networks (LANs)

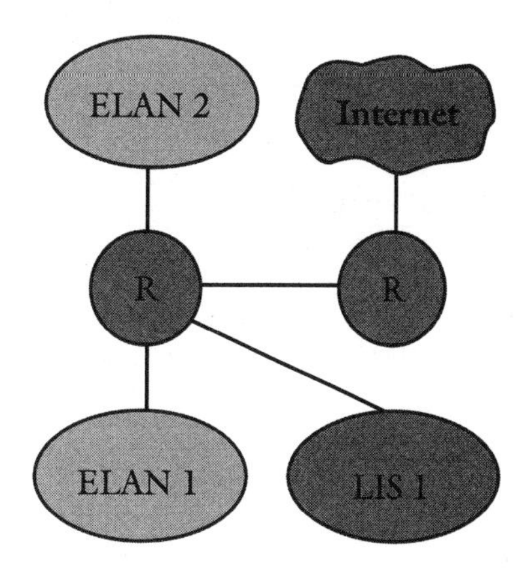

LANE was designed as migration protocol, allowing ATM-attached stations to communicate with each other using applications written for connectionless environments, as shown in Figure 5.55. LANE was also designed to support bridged communications between legacy stations over an ATM backbone. The intent was that this be a migration technology—employed only to allow for graceful switch to an ATM desktop solution with LAN applications rewritten to the native ATM API. The fact is, there are still no native ATM applications (other than LANE itself). LANE is now being remarketed (and redesigned) as an ATM-based, bridged VLAN solution.

FIGURE 5.55
LANE clients

LANE Clients

- LAN switches are proxy ATM-attached clients
- Endsystems are non-proxy ATM-attached clients
- **LANE is an ATM-based, bridged VLAN solution**

LAN Switch

ATM-25 Switch

Servers or PCs

ATM Forum Multiprotocol Over ATM (MPOA)

ATM MPOA (see Figure 5.56) is identical to LANE, except for the addition of cut-through routing service as an ATM network function, as shown in Figure 5.57.

The Data Protocols
RFC 1483 – Multiprotocol Encapsulation
RFC 1577 – Classical IP (CIP)
Next-Hop Resolution Protocol (NHRP)
ATM Forum LAN Emulation (LANE)
ATM Forum Multiprotocol Over ATM (MPOA)
IP Switching and Multiprotocol Label Switching (MPLS)

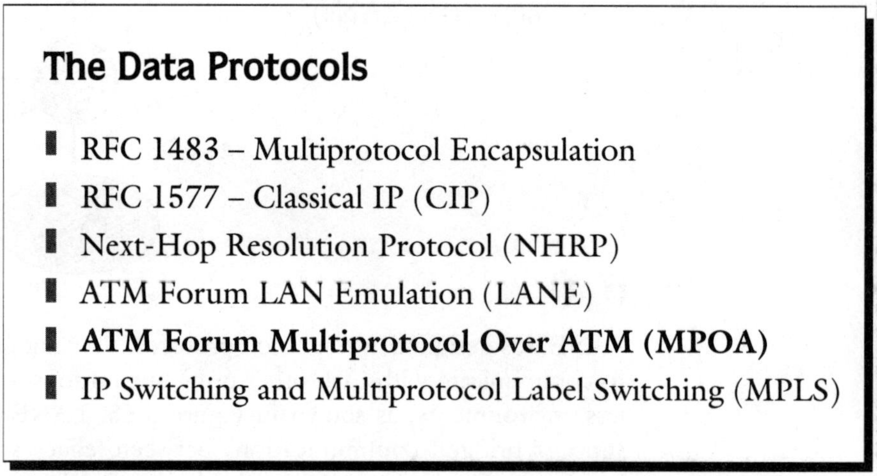

FIGURE 5.56 ATM Forum Multiprotocol Over ATM (MPOA)

MPOA
Signaling
Multicast
ATM
Addr reg
Cut-through
Internet
R
R
LAN Switch
ATM-25 Switch
Endsystems

FIGURE 5.57 MPOA's diagram

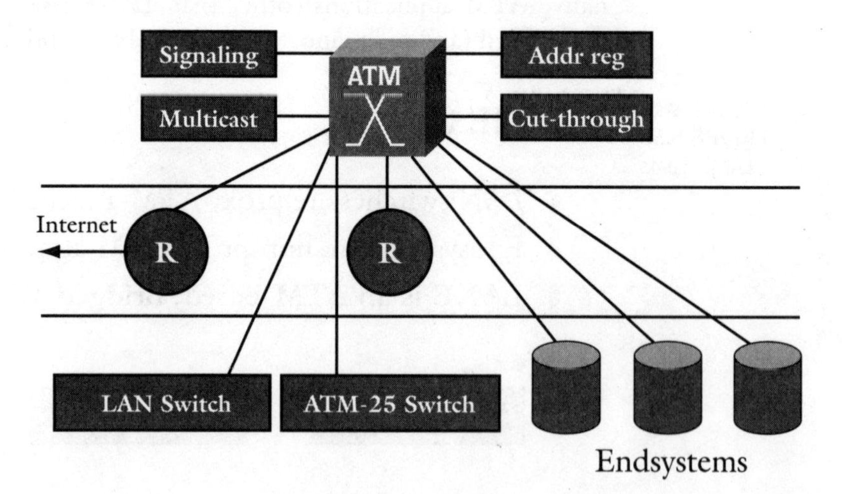

ATM MPOA services, shown in Figure 5.58, provide a means of mapping connectionless, multicast-capable LAN environments, and intranets into connection-oriented backbone environments. MPOA thus provides ATM services for resolving LAN MAC addresses and network addresses into ATM addresses, allowing MPOA clients to establish a connection to a destination station or host by using the ATM network's signaling service. MPOA also provides mechanisms for mapping LAN multicast into (tree-based) ATM multicast.

FIGURE 5.58
ATM MPOA services

MPOA Services

- AF-UNI 3.x/4.0
- ARP—Layers 2 and 3
- Multicast—Layer 2
- Route—For cut-through between VLANs and subnets

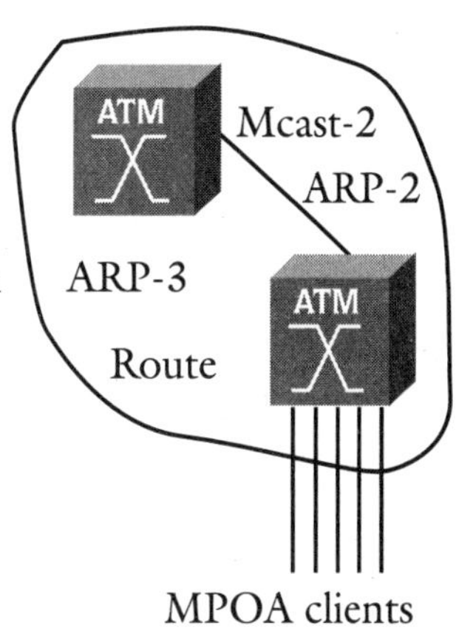

MPOA adds NHRP to the LANE one-armed router. In addition, there is a mechanism for distributing and caching shortcut (cut-through) addressing information to MPOA clients. MPOA clients then can perform Layer 3 packet forwarding functions and cut through the ATM network—avoiding router hops along the way.

Figure 5.59 shows how the one-armed router of LANE acts just like a traditional router, performing packet forwarding and route processing, but only with logical ports, instead of physical ports. MPOA adds NHRP function to these one-armed routers. With the addition of NHRP, the one-armed router can shed the responsibility for steady-state packet forwarding, which is now handled by clients directly, since NHRP allows for the establishment of shortcuts between MPOA clients. MPOA thus defines a distributed routing solution, separating routing function from packet forwarding function.

As for MPOA clients, just as NHRP enhances, actually replaces, CIP with shortcut routing, MPOA is LANE-enhanced with short-cut routing, as shown in Figure 5.60.

FIGURE 5.59
MPOA routing services

Routing Services

- GW to the Internet
- One-armed router now performs default forwarding of packets and route processing—routing information distributed to MPOA clients

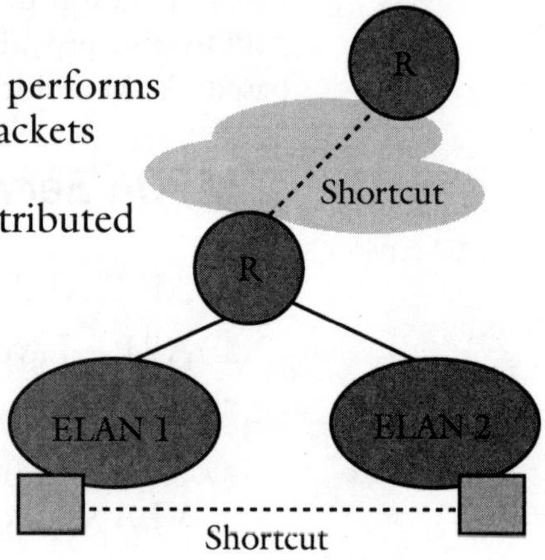

R
Shortcut
R
ELAN 1
ELAN 2
Shortcut

FIGURE 5.60
MPOA clients

MPOA Clients

- LAN switches are proxy ATM-attached clients
- Endsystems are non-proxy ATM-attached clients
- **MPOA adds network shortcuts (via NHRP) to LANE**

LAN Switch
ATM-25 Switch

IP Switching and Multiprotocol Label Switching (MPLS)

IP switching and MPLS (see Figure 5.61) require a single signaling service in the ATM backbone, as shown in Figure 5.62. Routers, LAN switches, servers, and workstations may be ATM-attached in this model as well.

In classic IP, LANE, and multiprotocol over ATM, MAC and IP addresses are mapped to ATM addresses, so that ATM signaling can be used. Because of that mapping, ARP services are required to resolve legacy addresses into ATM addresses (see Figure 5.63). In IP switching protocols, IP flow or topology is mapped to ATM virtual circuits directly. This obviates the mapping of legacy addresses to ATM addresses.

FIGURE 5.61
IP Switching and Multiprotocol Label Switching

The Data Protocols

- RFC 1483 – Multiprotocol Encapsulation
- RFC 1577 – Classical IP (CIP)
- Next-Hop Resolution Protocol (NHRP)
- ATM Forum LAN Emulation (LANE)
- ATM Forum Multiprotocol Over ATM (MPOA)
- **IP Switching and Multiprotocol Label Switching (MPLS)**

FIGURE 5.62
IP Switching and Multiprotocol Label Switching

IP Switching and MPLS

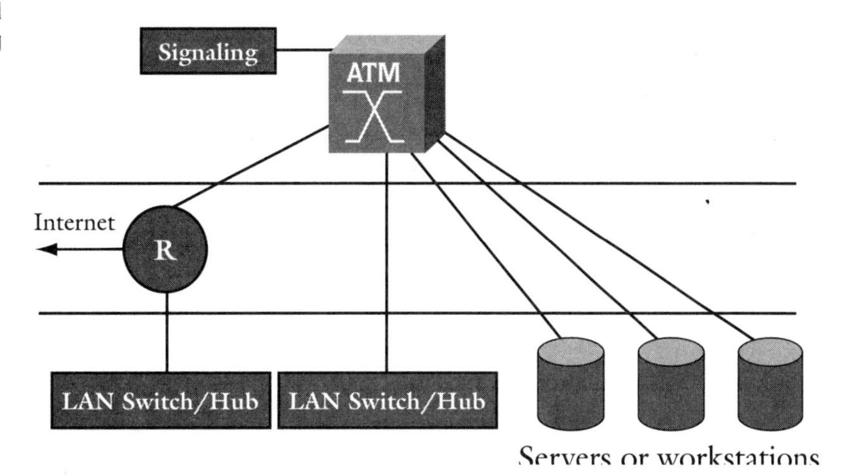

The purpose of IP switching is to avoid replication (see Figure 5.64). ATM provides new complex mechanisms in support of multicast, routing, and addressing. There are already protocols (and applications) supporting these functions for IP. IP switching marries IP functionality with the hardware benefits of ATM switching. The result is that IP routing function is applied at the edge of the ATM network and ATM switching becomes the high-speed packet-forwarding core.

FIGURE 5.63
ATM IP switching services

IP Switching Services

- IETF GSMP signaling and switch control
- No ATM Forum signaling
- IP flows and/or topology mapped to ATM connections

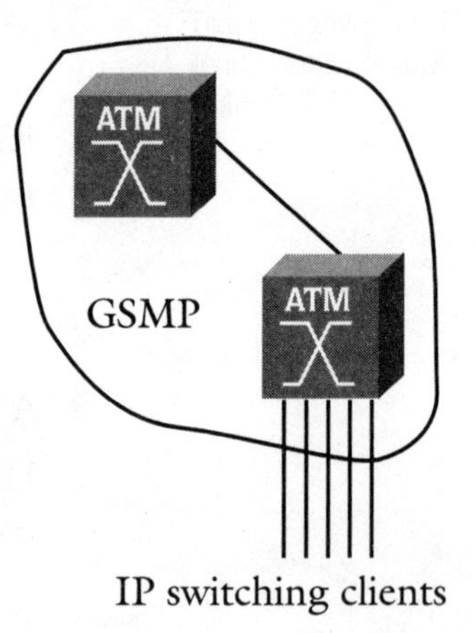

FIGURE 5.64
IP switching routing services

Routing Services

- Legacy clients can be behind ATM-attached routers
- Routing services performed by routers
- Packets only pass through routers at the edge

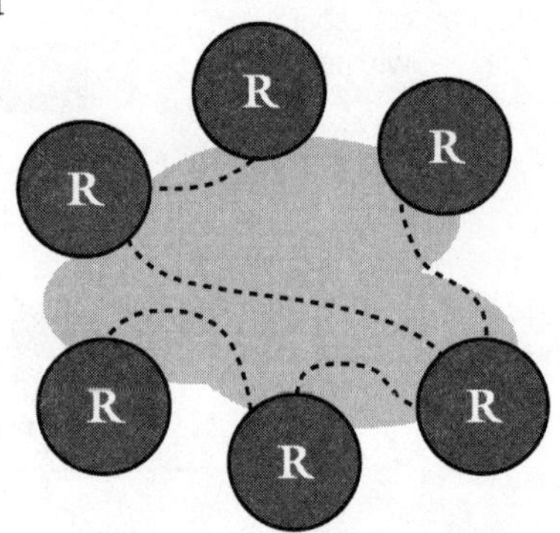

For the IP switching client, much as with NHRP and MPOA, the concept of IP switching is to avoid processing packets in routers in the middle of the ATM network—where possible router hops are avoided (see Figure 5.65).

FIGURE 5.65
IP switching clients

Clients

- Routers (packet forwarders) are IP switching clients
- LAN switches and endstations can be routers
- **IP switching is a solution for building router-based, ATM-switched backbones**

LAN Switch LAN Switch

Endstations

Summary

In summary, LANE, MPOA and IP switching provide solutions which allow for building a robust backbone network based on an ATM switching core. Figure 5.66 provides an outline of the features offered by IP over ATM.

FIGURE 5.66
Features and standards of IP over ATM

Conclusions

- RFC 1483 is an encapsulation method—it enables ATM-attaching routers by making VCs look like point-to-point links
- CIP is a high-speed workgroup solution
- NHRP replaces CIP and adds shortcut routing
- LANE, MPOA, and IP switching/MPLS are data network solutions

What's Next

This chapter discussed the ATM data model, its network services, data protocols and LANE, as well as ATM MPOA services. The next chapter discusses digitization methods and packet-based voice applications.

CHAPTER 6

Broadband Packet Networks and Voice Communication

Communications systems evolve to meet the needs of the computing environment. We are in the midst of a major IT evolution, actually the second revolution in the computing environment, in that we are moving from the "time-shared" host-based network to an internetwork of high-powered workstations.

The first revolution of computing was the evolution from batch to interactive processing, prompted by a dramatic decrease in the price of computing equipment. The mode of scheduling the work of the user around the computer (batch) evolved to scheduling the work of the computer around the user (interactive).

This move to internetworked workstations, however, has not removed the need for communications, although it has vastly altered the fundamental characteristics of these communications. The traditional host-based system is based on the transfer of information in relatively small units. In this mode, no more information is normally transferred than the amount of information that will fit on a single screen—about 1920 characters (24 lines of 80 characters each).

This interactive processing has become our dominant mode of computing over the past ten to fifteen years. Consequently, our current data communications networks have been designed to support this particular type of networking environment, as shown in Figure 6.1.

FIGURE 6.1 Traditional host-based network

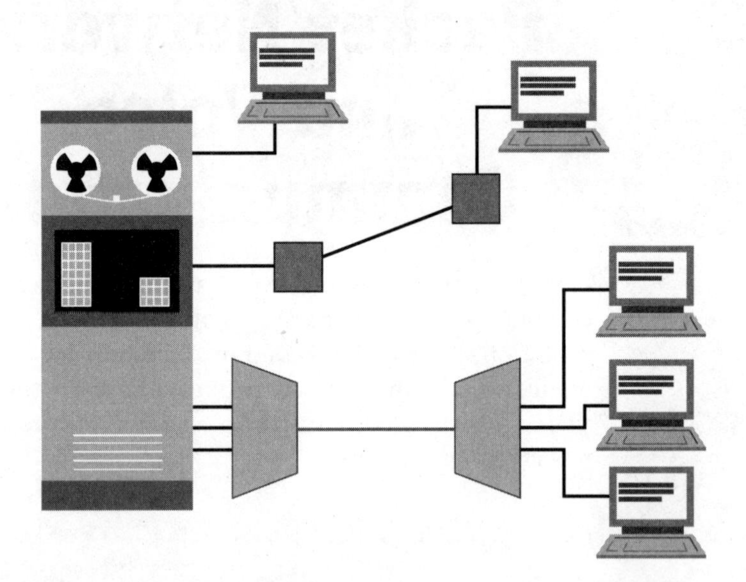

This chapter discusses the traditional host-based network model and its evolution to broadband packet networks. One of the major players in the broadband packet networks is Nuera Communications, which has a significant presence in the VoIP industry. Nuera Communications is a market leader in advanced voice communications equipment. It was founded to provide technology and products unmatched in the areas of telephony, call processing, and voice compression.

This section was based on a white paper written by Steven Taylor, of Nuera Communications. For a full version of this paper or for additional information about Nuera's VoIP technology and products, see Chapter 8 or visit Nuera's Web site at **URL http://www.nuera.com**.

Broadband Packet Networks

Figure 6.1 shows how, traditional host computers, from minicomputer to mainframe to supercomputers, are linked directly to other computers or terminals, even though the actual line first goes into a front-end processor, typical for a minicomputer. The typical speed for the communications interfaces is 64 kbps or less.

The most typical connection is the local one, a terminal directly connected to the host represented in Figure 6.1 as the terminal/computer directly connected to the host. Data communications techniques were developed to allow remote terminals to operate as if locally attached. The data transmission speed need for each terminal is limited in reality by the "screen-based" mode of communications. The user digests the contents of one screen of information before requesting another screen. Increasing the speed of the transmission line increases the speed with which the screen is repainted, but provides few other operational advantages. Thus there is little to be gained once the screen is repainted very quickly.

The terminal/computer connected through the squared box in Figure 6.1 has a link to the host computer. This is a link to a single remote device, usually over an analog telephone line, with typical speeds up to 19.2 kbps. If a faster link is needed, this may also be accomplished at a typical speed of 64 kbps via ISDN. In the case of analog telephone lines, the squared boxes represent modems, but in the case of ISDN they represent terminal adapters. This type of connection is usually a switched connection. The connection is established for each session, and connectivity is possible to a number of points via some form of "dialing."

If there is a need to connect a number of terminals using a single communications link, a multiplexer would be required. The trapezoidal representations in Figure 6.1 represent multiplexers performing the same basic function, allowing multiple conversations to share a single transmission facility, whether the transmission line is as simple as 9.6 kbps modems with an analog phone line or as complex as a 45 Mbps T3 service. The multiplexers come in two basic flavors: time division and statistical.

Notice the terminals in the remote clusters attached to the multiplexer. The speed for each interface is usually the same speed as the computer interface. The type of multiplexer used and the traffic characteristics will determine the maximum data throughput per terminal. The data transmission speed requirement for each terminal is limited in reality by the "screen-based" mode of communications discussed earlier.

The Evolution to Workstation-Based Systems

The move to workstation-based computing has evolved as the power of the personal computer has risen exponentially. Many users now have more computing power on their desktops than mainframe computers had just a few years ago. This vast increase in power has moved many of the computing tasks from the mainframe to the personal workstation. This has not, however, resulted, in a decreased need for communications.

It has resulted in vast changes in the mode of communications. Now, rather than sending information in a screen-by-screen fashion based on terminal-host communications, we send information in a file-by-file mode, as shown in Figure 6.2. Also, while the quantity of this information is vast, the rate of the transmission is much more sporadic. In some cases, as many transmissions per minute may be needed as in terminal-host communications. This might be the case if the application used high-resolution graphics with a supercomputer acting as the host and the high-performance workstation acting as the terminal. In other cases, a database may be downloaded, massaged for minutes to hours, then uploaded to another system.

FIGURE 6.2 Workstation-based computing generates file-by-file-based communications

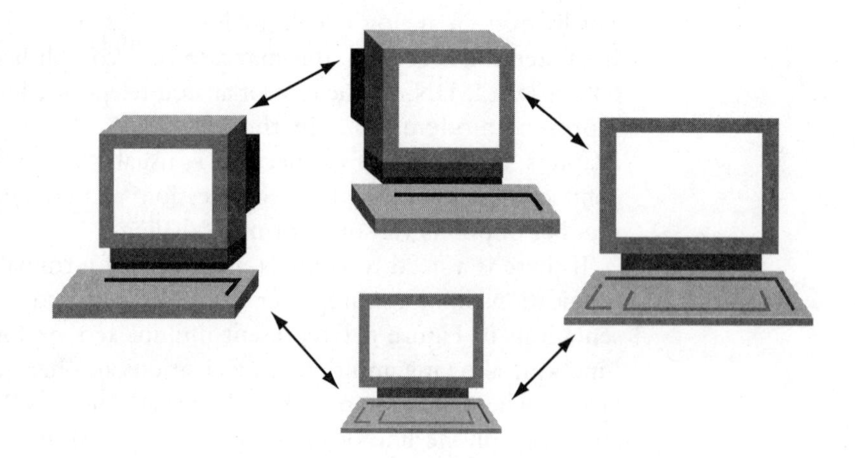

The term **file** used as base for workstation-based communication is used here in the context of a large amount of information consisting of many more than 1,920 characters. Depending on the actual application, this file may constitute an actual data file, such as a part of a database; a video image; or any other collection of data to be shared.

These applications have been under development for several years using intensive data sharing on the LAN. Literally several million bits per second of information throughput are available between workstations at a relatively low cost. In the wide area network, though, this is not the case. While multimegabit-per-second throughput is available, it is still quite expensive. The challenge is to provide connectivity across the WAN for highly bursty high-bandwidth applications to support this change in the norm for the computing environment. Meeting this challenge is the goal of broadband packet networking.

Broadband Packet for LANs

When any of the broadband packet technologies are discussed, the first and foremost application mentioned is the LAN internetworking—the interconnection of remote LANs. The genuine need is not because there are LANs that require interconnection, but because the type of traffic typical of LANs now needs to be carried across the WAN.

As discussed earlier in this chapter, file-based, as opposed to screen-based, communications tend to be quite bursty. The nature of these applications varies from file servers to shared graphic images, and file-based applications are already becoming commonplace in LANs. Now demand is growing that these same tasks be performed across the wide area. We are thus faced with supporting bursty, high-bandwidth applications across the wide area while being constrained by relatively expensive bandwidth across the same area.

We must be careful to remember, though, that even though LAN interconnection is discussed as the killer application, because the type of traffic that needs broadband packet technology is usually associated with LANs. Host-to-host traffic may generate this type of traffic as well, depending on the application, and the presence or absence of actual LANs is inconsequential. Conversely, LAN-to-LAN traffic, as shown in Figure 6.3 does not really need broadband packet if the traffic is terminal-to-host traffic that just happens to be using a LAN as the local transport mechanism.

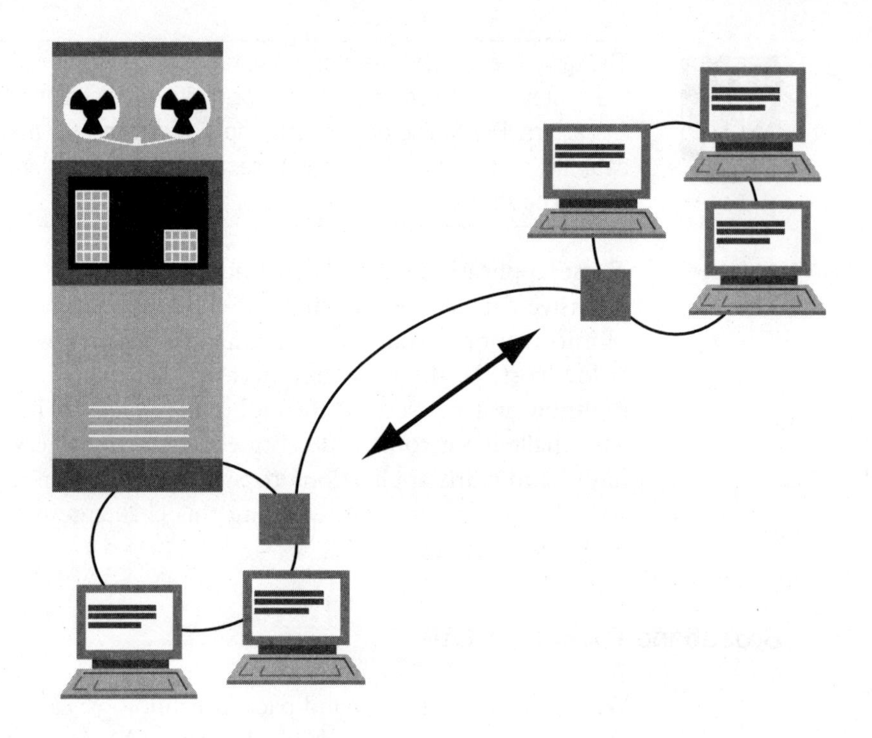

FIGURE 6.3 Typical LAN-to-LAN traffic does not need broadband packet if traffic is terminal-to-host

Internetworking LANs

Internetworking, the interconnection of LANs via bridges and routers, has become standard practice and is one of the fastest-growing areas of the networking marketplace. The challenge of interconnecting LANs using bridges and routers comes in choosing appropriate transmission facilities in the wide area portion of the network. LANs typically run at several megabits per second; information is transferred among devices connected to the LAN at very high rates of speed.

When LAN internetworking is implemented, the intent is to allow geographically dispersed devices to perform as if they were all connected to a single LAN. While this is possible insofar as providing physical connectivity and the potential for information transfer, the actual rate of information transfer is usually vastly inferior to that on the LAN itself. The connectivity is there, but the performance may not be there, especially for file-based interactions.

The reason for this lack of performance is that LANs with megabits per second of transmission speed are interconnected via WAN transmission facilities with tens of kilobits per second. This model works well for transaction-based communications based on a terminal-host communications model.

However the vast mismatch of speeds makes LAN-WAN interconnections insufficient for the emerging file-based communications applications.

When file-based transactions are needed, there are two choices: use traditional speeds in the 56-kbps range with relatively poor performance but good utilization of the facilities, or move to T1 which provides excellent performance but only uses the available bandwidth a small percentage of the time. It is exactly this dilemma that broadband packet networks address.

Figure 6.4 illustrates a typical internetworking of LANs. The larger shaded squares on the figure are bridges or routers used to provide connectivity among LANs via wide area transmission facilities. In this case, the transmission facilities are dedicated. Each LAN (A, B and C) can have its own physical topology, varying from token ring LAN to Ethernet, either coax or twisted pair, or even an FDDI ring. The only key is support by the bridge or router.

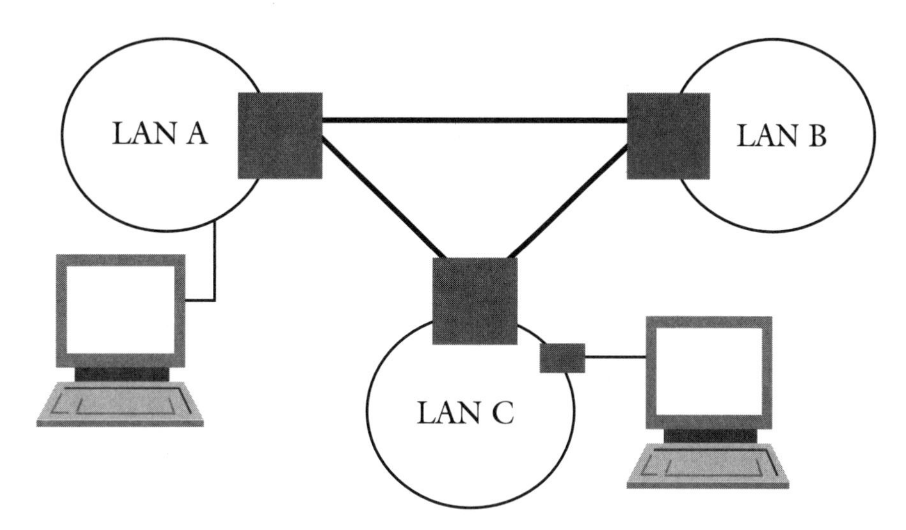

FIGURE 6.4 Typical LAN internetworking

The smaller rectangles connecting the terminal/computer to the LAN represent media access devices. These may be an external devices, such as terminal servers, or internal to the computer, workstation, or other computing device. If this is a token ring network, the media access device may be called a MAU (Media Access Unit), for example.

Packetized LAN Interconnection

When LANs are interconnected via some form of packetized network, network bandwidth is allocated only when it is actually needed. These networks become especially appropriate for interconnecting file-based LAN-to-LAN communications when the network is a broadband packet network, as show in Figure 6.5.

The term "broadband packet" is used to indicate that the speeds supported by the packet networks are in the range of megabits per second and higher. Several megabits per second may be allocated to a particular file transfer for the seconds that the bandwidth is needed. Rather than allocating tens of kilobits per second (e.g. 56 kbps) on a constant basis, a much larger amount of bandwidth (e.g. 1.544 Mbps) may be allocated, but only for the amount of time that it is actually needed. At any other time, other users may vie for this pool of dynamically assigned bandwidth. Indeed, this may be considered a form of fractional T1, as discussed later in this chapter.

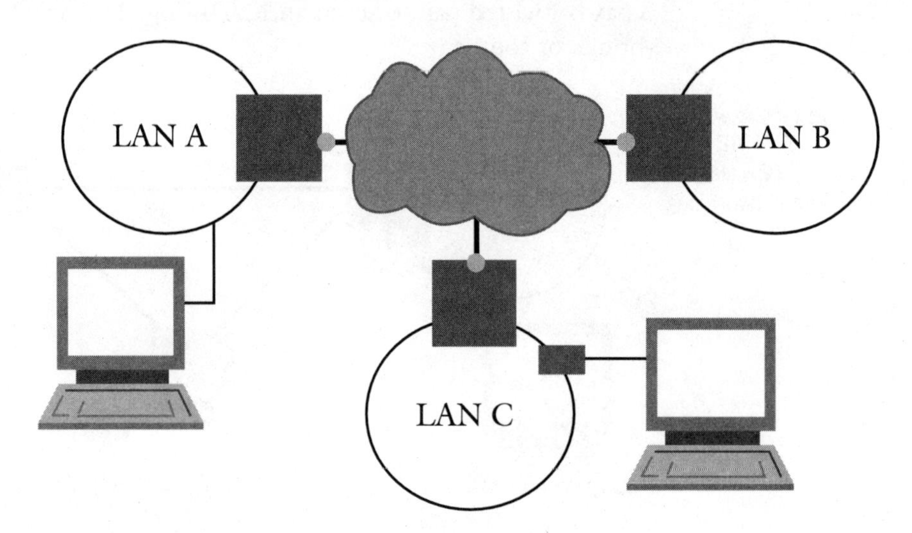

FIGURE 6.5 Packetized LAN interconnection

In Figure 6.5 the larger squares represent bridges or routers used to provide connectivity among LANs via wide-area transmission facilities. In this case, the bridge/router supports an interface designed for use with a broadband packet network of some type (represented by the small circles).

These small circles connecting the larger squares to the Internet cloud are interfaces for the type of broadband packet network being used, frame relay, SMDS, or ATM. The specifics for each of these interfaces will be discussed in detail later in this chapter. These interfaces guarantee that the data is packetized in an appropriate format for transport by the network.

The physical interface will typically be a standard serial interface of some form. At speeds up to a few megabits per second, this will typically be a V.35 interface. At higher speeds, the HSSI interface is typically found. The type of service (frame relay, SMDS, or ATM) and the nature of the service (public or private network) will determine the details of the connections.

The broadband packet network is the heart of the system. It is designed to carry traffic in a packetized (statistically multiplexed) format and to deliver it to a similar interface at a remote location. Depending on whether a private network is implemented or a public network service is used, the actual packet switching equipment may be within the carrier network or may be located on the customer's premises.

The key to the packet network is *lots of traffic* from *lots of sources.* This allows the bandwidth that might otherwise be dedicated to individual point-to-point dedicated connections to be aggregated and parceled out to the different connections on an as-needed, dynamic basis.

Understanding Fractional T1

Fractional services are services in which a portion of a whole is purchased. Currently, when we hear of services like fractional T1, we usually think of the service as a dedicated bandwidth portion of a larger dedicated bandwidth service. For instance, as depicted in Figure 6.6, one may buy 256 kbps or 512 kbps instead of the entire T1 (1.544 Mbps). In most cases, the bandwidth purchased will be an integral multiple of either 56 kbps or 64 kbps.

FIGURE 6.6
Traditional bandwidth-fractional T1

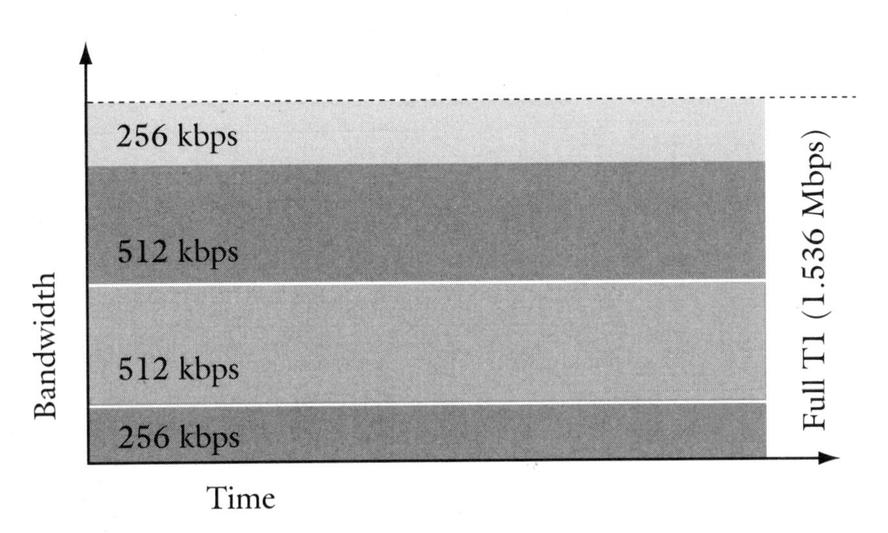

When the bandwidth is purchased in this fashion, it is called "bandwidth-fractional" because a fraction of the full bandwidths bought over a relatively long period of time, weeks to months to years.

Time Fractional T1/T3

It is also possible to purchase transmission capabilities by a method by buying all of the bandwidth, but only for a packet time. No carriers really charge for the packet time in those exact terms, but that is exactly what charges pay for whenever a packetized service is used.

The same principle applies when using any packetized service, on a public or a private network. Since there is access to all of the bandwidth for a fraction of the time, the term "time-fractional" is used, to distinguish between the two possible types of fractional services (see Figure 6.7).

FIGURE 6.7
Time Fractional T1

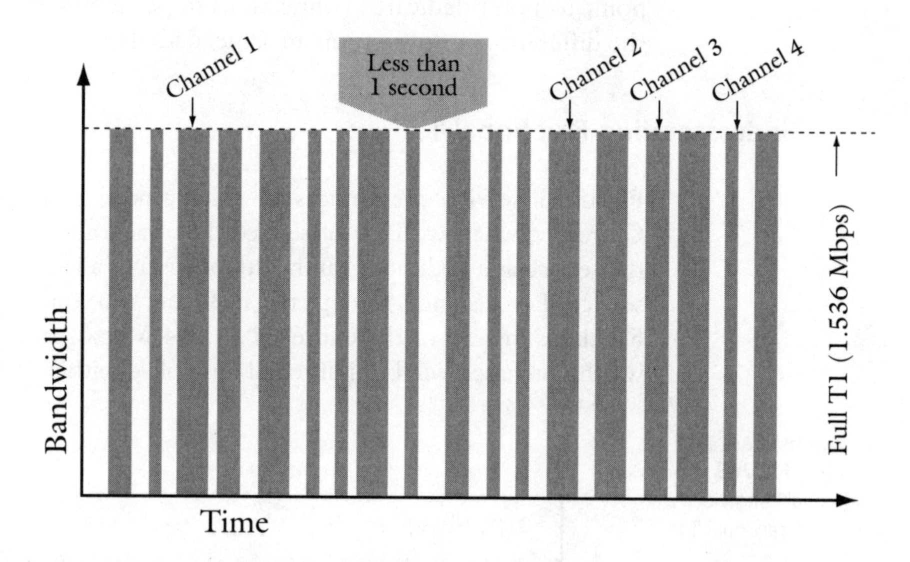

Time-fractional services, such as those provided by the broadband-packet technologies, are an excellent fit for the evolving needs for file-based communications. If a file of several megabytes of information needs to be transmitted, all of the available bandwidth is needed for the duration of the transmission. Once the file transmission is finished, no bandwidth is needed. Thus, the time-fractional model provides an excellent complement to the current and emerging needs for wide area communications, especially multimedia applications, such as VoIP.

Multiplexers and Framing

Bandwidth-fractional and time-fractional services are really just another way of looking at multiplexing. If a single organization has access to the entire transmission facility, such as a T1 circuit, and it subdivides the transmission facility among several tasks, this is the classic use of multiplexers in the private network. Carriers have used similar techniques for years to allocate bandwidth to multiple users.

Any type of digital multiplexer requires some form of framing. The signal coming from the transmission facilities is nothing more than a series of ones and zeros. There is nothing about these ones and zeros that inherently separates one conversation from another. Thus, there is a need for some form of structure to identify which ones (sheep) and zeros (cows) belong to which of the multiple conversations being transported, as shown in Figure 6.8.

FIGURE 6.8 Multiplexer and frames: identifying different types of data

TIP

What's in a name? What is a digital multiplexer?

A digital multiplexer is a multiplexer used to combine various types of traffic to be transported via digital facilities, such as 56 kbps or T1 facilities.

We noted earlier that there are two fundamental types of multiplexers: circuit, or time division, multiplexers and packet, or statistical, multiplexers.

Circuit Multiplexing

Circuit (time-division) multiplexing, is the most basic form of digital multiplexing. The time-division multiplexer divides the available bandwidth among the different applications on a dedicated basis. Each application has its share, and one application may not "borrow" from any other applications based on whether the bandwidth is active (in use) or not, as shown in Figure 6.9.

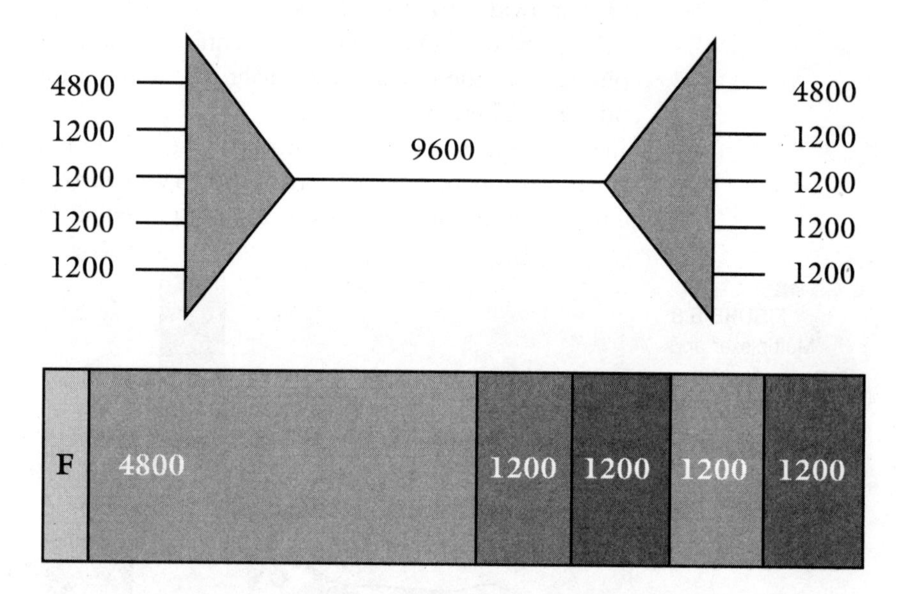

FIGURE 6.9 Time-division multiplexing

The dedicated nature of the bandwidth has several key characteristics:

- ✔ Circuit multiplexers tend to be relatively fast and relatively inexpensive. Since they are conceptually quite simple, circuit multiplexers are the first multiplexers to be introduced as transmission speeds go higher and higher.
- ✔ The dedicated bandwidth guarantees transparent transmission of data. The multiplexer ships bits from one end to the other without regard for protocol. Thus, by definition, virtually any type of traffic may be transported by a circuit multiplexer.
- ✔ There is virtually no variation in the delay from one set of bits to the next.[1] This is a function of the transparent nature of the device. Variations in delay usually result from delays due to traffic congestion and retransmission. These delays are found in packet multiplexers. Consistency in delay obviates the problems caused by variable delays when transporting synchronous protocols like SNA and X.25.
- ✔ Circuit multiplexers only require minimal framing overhead.[2] Most implementations of circuit multiplexers require less than one percent of the total bandwidth for this type of overhead. For instance, the stan-

1 There is indeed some variation speeds, commonly referred to as "jitter" and "wander." However, the variations from these factors, as well as from other factors like Doppler shifts in satellite systems, are microscopic and insignificant for the discussions in this section.

dard channel bank[3] uses only one out of 193 bits in the frame for framing in the standard SuperFrame[4] (SF) format. In the extended super frame[5] (ESF) format for channel banks, the framing actually used to identify channels consists of 1 out of 772 bits.

In Figure 6.9, the numbers aligned vertically on the left and right sides of the figure represent the various inputs to the multiplexer. The channels may carry any type of traffic and the speeds for the channels are arbitrary. They may be any mix, but, since the bandwidth is dedicated, the sum of the speeds must not exceed 9600. In fact, the sum of the speeds must be slightly less than 9600, so the figure does not provide an accurate account of this process. However, the framing overhead for time division multiplexers is minimal.

The triangles represent multiplexer hardware. The multiple-channel side is the input/output for the various channels, and the single-connection side represents the connection to the transmission facilities. For simplicity, all of the multiplexers shown in Figure 6.9 are point-to-point single-transmission link applications. In reality, many multiplexers, both circuit and packet, support switching capabilities and multiple links for more complex network applications.

In Figure 6.9, the amount "9600" represents the transmission facilities between the two multiplexers. Use of the speed "9600" is for demonstration purposes only. This may be thought of as a 9600-bps modem link. It might equally as well have been a 1.544-Mbps (T1) link connecting two channel banks. In that case, the link would have supported twenty-four 64-kbps circuits.

The numbered portions of the frame diagram represent the "payload" of the frame. This is the information that is transported for each individual channel. Note that these are fixed information payloads for each channel, resulting in a dedicated bandwidth for each channel.

---

2 Framing overhead is the amount of overhead required to perform the framing functions, that is, the overhead used to keep track of which bits belong to which conversation.

3 This is a type of digital multiplexer that has become a commodity within telephone companies. The channel bank transforms 24 analog voice conversations to 64 kbps digital voice (per conversation) and formats the digital conversations for transport across a single 1.544 Mbps transmission facility (T1 line).

4 SuperFrame is a framing format used in "D3" and "D4" multiplexers. It consists of a "superframe" of 12 frames. Each of the 12 frames contains a single 8-bit sample from each of the 24 channels, plus one framing bit.

5 Generally viewed as an enhancement for SF framing, the ESF format reassigns the use of the single framing bit in each 193-bit frame. A quarter of the "framing bits" are used for identifying the channels, a quarter are used to provide error detection (for diagnostic purposes only, not for retransmission), and half are used for supervisory communications and control.

Packet Multiplexing

Packet multiplexers, often called statistical multiplexers, are an alternative means for transporting data from one point to the next. The packet multiplexer derives its power from transporting the data in packets (message units) rather than as a continuous flow of information.

Most data are not continuous but usually occur in bursts of some form. When there is a burst of data, bandwidth is assigned to transport that data across the WAN. When there are no data to be transmitted, no wide-area bandwidth is assigned. The packet multiplexer differs from the circuit multiplexer in that there is no preassigned dedicated bandwidth for any of the channels. Bandwidth is assigned on demand, as shown in Figure 6.10.

FIGURE 6.10 Characteristics of packet multiplexers

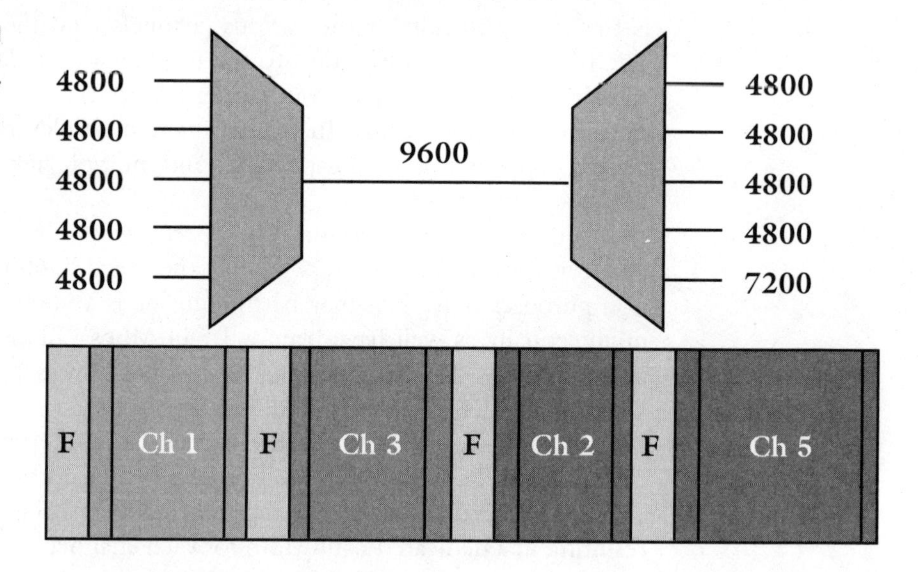

This capability lends distinct characteristics to the packet multiplexer, as shown in Figure 6.10:

- ✔ Protocol sensitivity is a requirement for any type of packet multiplexer. Since the multiplexer only transmits data when there are "real" data to be transmitted, it must be able to determine the difference between "real" data and traffic that may appear on the line as fill. It is the removal of the "fill" traffic prior to transmission and reinsertion on the opposite end that provides the vast majority of the efficiency found in packet multiplexers. Still, the multiplexer must have some knowledge of the protocol in order to recognize and remove the fill characters.

- ✔ Packet multiplexers can provide extremely efficient use of the bandwidth available on the transmission facilities. Since no channels have any bandwidth dedicated for their exclusive use, the bandwidth is available to channels on an as-needed basis. If the total traffic pattern has many channels and each channel needs the bandwidth on a "bursty" basis, the aggregate transmission bandwidth may be shared.
- ✔ Since the aggregate transmission bandwidth is shared among the channels, the delays in the packet system will be much more variable than those in the circuit multiplexer. If there are only a few channels contending for the bandwidth, the amount of bandwidth available to each channel is fairly high. As more and more channels contend for a fixed amount of bandwidth, the percentage of the total available to each channel is lower, and the overall delay increases. (The transmission bandwidth between the two packet multiplexers is still limited to the speed of the physical facilities.)
- ✔ By necessity, packet multiplexers have higher framing overhead than do circuit multiplexers. This is because the data in each "packet" of information must be explicitly addressed or labeled. In the circuit multiplexer, the channels are identified by their position in the frame relative to the framing bits. In the packet multiplexer, there are no predefined framing bits. This requires a label on each packet of information to identify its channel.
- ✔ Sometimes packet multiplexers have a protocol of their own used between multiplexers to guarantee accurate delivery of data. This is especially a hallmark of packet switches and statistical multiplexers, devices that were developed primarily to transport asynchronous ASCII data that has no inherent protocol. While the retransmission of erred packets guarantees accurate delivery, it also increases the delay variability because packets that are not retransmitted are delivered more quickly than packets that require retransmission. The addition of this layer of protocol also adds to the processing power needed in the multiplexer.

In Figure 6.10, the column of 4800 numbers on the left represents the input/output channels for the packet multiplexer. Note that the sum of the channel speeds can exceed the speed of the transmission links. It is not possible, however, for the channels to operate continuously and simultaneously at these speeds. Rather, these are the speeds at which data are communicated to and from the multiplexer. The actual throughput rate will depend on the traffic at the time.

The rate at which the various channels submit data to the multiplexer may actually exceed the transmission rate for a short period of time. The excess data will be stored temporarily in buffers in the multiplexers until the traffic has dropped off sufficiently to empty the buffers.

The actual throughput for an individual channel will never exceed the rate of the transmission facilities. In fact, it will always be somewhat less due to overhead, even if the channel has sole access to the facilities.

The list of channels across Figure 6.10 represents the data that will be specific to the data from a given channel. For simplicity, the area shown here includes both the "packet" and the "frame" portions of protocols like X.25, so the use of the term "packet" here indicates that the data are transmitted in discrete bundles or message units. It does not refer explicitly to the packet level of the X.25 protocol.

The hashed area at the beginning (F) is the header. This will contain information that identifies the contents as belonging to one particular channel. The solid area in the center will contain the actual data from a specific channel. If the packet multiplexer is "frame-oriented," the actual length (number of bytes) in the payload will be variable. If the packet multiplexer is "cell-oriented", there will be a fixed number of bytes in the payload. The hashed area at the end represents any "trailer" information that might be present, depending on the actual format used. In many cases, the trailer information will contain a 16-bit error detection code (CRC—Cyclic Redundancy Check) that is used to determine whether an error occurred during transmission.

Packet Switching

Packet switching is a logical extension of packet multiplexing. Each message unit (frame or cell) in the packet-multiplexed data stream has a unique address identifying which channel "owns" that particular information. The packet switch can accept this information and route it to an appropriate next destination based on the information identifying the "owner," "owner."

As depicted in Figure 6.11, each message unit entering the switch has a message unit identifier—in this case either an "A" or a "B"—identifying the owner of the information. Assuming that "A" and "B" are at separate locations, the switch then sends that information along an appropriate path to either the "A" or "B" destination.

Be careful not to take the "A" and "B" designations too literally in this example. Depending on the design of the system, the "A" or "B" identifier may actually stay with the message throughout the transmission path (as depicted here), in which case the "A" or "B" is a form of global address, much like the address on a letter. This is a major characteristic of a connectionless architecture.

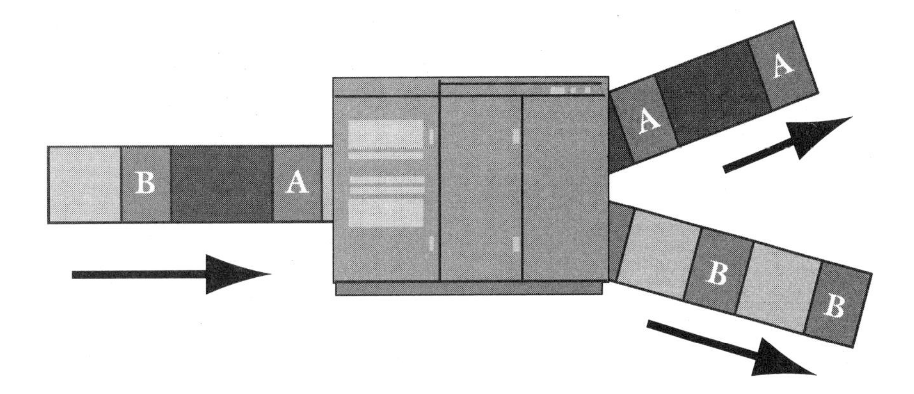

FIGURE 6.11
Packet switching

Conversely, the switch may keep track of the connections and use the "A" or "B" as a circuit number for only a singular transmission facility. In this case, a single message may actually have several different "message identifiers" as it traverses the network. This method allows the addresses to be reused on every link, supporting a larger network for the same length address. This type of network is said to be connection-oriented.

Connection-oriented architecture is not inherently better than connectionless. Each has its own strengths and weaknesses; so do not be misled by claims that one is "good" and the other is "bad." Many of the advantages and disadvantages of each are more important to equipment designers and network providers than to end-users.

Of the broadband packet technologies two—frame relay and ATM—are connection-oriented. The third, SMDS, is connectionless.

Understanding Broadband Packet

Broadband packet switching is a specialized form of packet switching and multiplexing. The use of the word "broadband" implies that the technologies are appropriate to be used at "broadband speeds," that is, at T1 (1.544 Mbps) and above.

Broadband packets are fundamentally the same as any other type of packets. They consist of a header, a payload, and an optional "trailer." All three[6] types of broadband packet networks have two fundamental assumptions in common, as shown in Figure 6.12:

6 Frame relay, SMDS, and ATM are the three technologies usually included under the "broadband packet" umbrella.

- ✔ Broadband packet networks are optimized for transporting protocol-oriented[7] traffic. The inherent protocol in the traffic will guarantee delivery, so the network need not perform that task. This allows the network to bypass processing tasks that guarantee delivery, thus accelerating the throughput possible with a given amount of processing power.
- ✔ Broadband packet networks assume relatively clean[8] transmission facilities. These greatly reduce the chances of encountering a transmission error. Error detection on an end-to-end basis performed inherently in the transported protocol becomes feasible.

FIGURE 6.12
Generic packet format

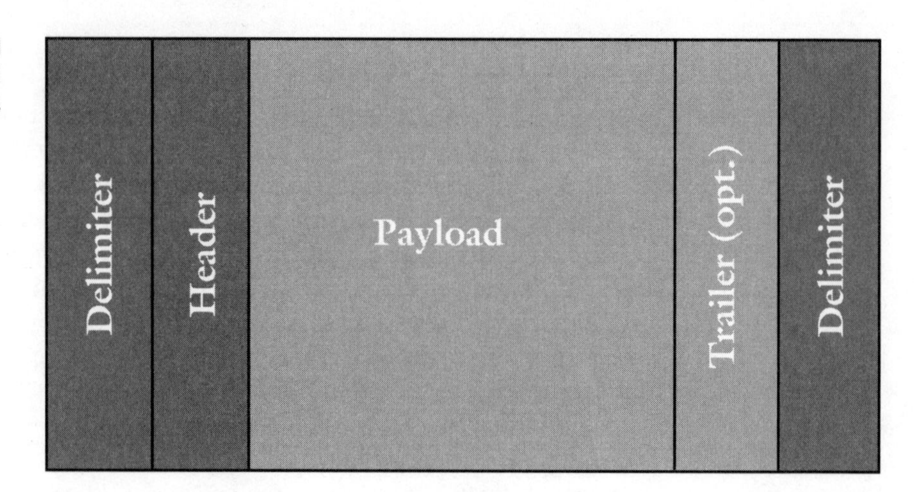

In Figure 6.12:

- ✔ The **flag** or **delimiter** identifies the beginning of a message unit. It separates one message unit from another. In frame-oriented technologies, the flag will usually be an 8-bit character with the hexadecimal value of 7E. Flags are also used to provide fill between frames. In cell-oriented systems, the function of the flag may be accomplished with a fixed format, much like the framing bit in the circuit (time division) multiplexer.
- ✔ The **header** portion of the message unit identifies the owner of the information. In frame relay and ATM, the header is used to identify the circuit

---

7 This is traffic being transported that already has a protocol, like SNA/SDLC, X.25, and most LAN protocols. In contrast, traditional asynchronous ASCII communications have no protocol, so the network, typically X.25 or statistical multiplexers, must protect against transmission errors.

8 These are transmission facilities with low error rates, particularly typical of fiber optic transmission facilities.

number on the individual link. The switches then provide connections among "circuit numbers" to route message units to their appropriate destinations. In SMDS, there are actually two forms of "message units." In one, circuit numbers are used just as in frame relay and SMDS. In the other type of message unit, the header actually contains the global address for the message unit. The header may also contain information to be used for congestion management and for error detection for control messages. Error detection in the header is usually found in cell headers. Frames usually use the "trailer" for error detection.

- ✔ The **payload** is the "information" portion of the message unit. It consists of the traffic being transported in its native format. A single payload may contain an entire native-mode message unit or a portion of a native-mode message unit. While the payload has no error detection functions within the broadband packet network, the payload will often contain error detection information as part of the upper-layer protocol being transported.
- ✔ Notice that **trailers** are optional. Some broadband packet formats use trailer information, others do not. When a trailer is used, its primary function is to provide error detection for the header and the payload. A trailer provides error detection for the entire payload, because of the convenience of using readily available hardware when building equipment, not to provide retransmission on error. This task is still left to the higher-protocol layers. The critical task of error control in a broadband packet system is to insure the integrity of control messages and to make sure that the header was not corrupted. This may be accomplished by placing error control either in the header itself or in the trailer.

The result of the above is that all three broadband packet implementations are much more similar than they are different, especially when compared with traditional technologies like T1 multiplexers and packet switches, offering:

- ✔ Bandwidth on demand (packet multiplexing or time-fractional service review
- ✔ Support for speeds of T1 and above, with ATM technology approaching the Gbps[9] range
- ✔ Connectivity to a large number of users with dynamic bandwidth assignment to all points (packet multiplexing): resulting in
- ✔ Excellent transmission characteristics for bursty, high-bandwidth data, as typically found in LAN internetworking applications.

---

9 Gigabits (a billion bits) per second.

The characteristics of the performance of broadband packet systems at high speeds lead to the term "fast packet" sometimes applied to the technologies. At the same time, none of the technologies offers guaranteed delivery of data, including retransmission on error. This is left to the higher-level protocols implemented in the DTE (This may be the LAN protocol or the protocol in the computer, such as SNA).

Since broadband packets come in various forms, such as frames and cells, it is convenient to have a term to use when referring to a generic packet. One of the most useful terms in this context is the PDU (Protocol Data Unit).

Frames and Cells

The basic difference between frames and cells is incredibly simple: frames have variable-length payloads and cells have fixed-length payloads.

Among the broadband-packet technologies, frame relay and one level of the SMDS interface protocol are based on frames. ATM and the other level of the SMDS interface protocol are based on cells.

The fixed versus variable lengths for frames and cells result in some distinguishing features, as shown in Figure 6.13:

- ✔ The delimiter between frames is a character (7E), and this character is recognized as a distinct portion of the data stream. By contrast, since cells are of a fixed length, the delimiter may be quite similar to the framing bit in the circuit multiplexer format.
- ✔ Frames generally have a "trailing" error control field. Cells generally include error control in the header. This is a function of convenience for equipment builders more than a fundamental indication of the usefulness of technology.
- ✔ Due to their simplicity, cell-oriented systems are generally viewed as more appropriate for higher speeds.
- ✔ The delay for cells is generally of shorter duration and is more predictable than that for frames, making cells generally more appropriate for voice and video. (This only matters if voice and/or video will be carried in the network.)

A cell is composed of:

- ✔ An **optional** delimiter used to separate cells. It will normally be provided from a fixed framing structure, similar (or even identical) to the framing bits in the circuit multiplexer.

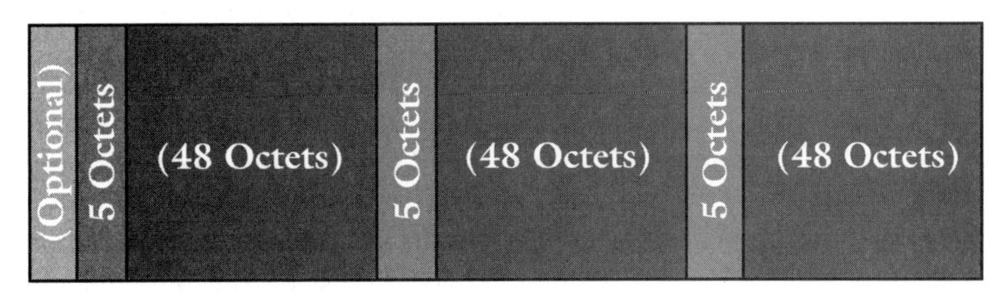

FIGURE 6.13 Cells have a fixed payload length while frames do not

- **5 octets**, the header portion of the message unit; these provide the identification of the owner of the information. In frame relay and ATM, the header is used to identify the circuit number on the individual link. The switches then provide connections among "circuit numbers" to route message units to their appropriate destinations. We have discussed the two forms of "message units" in SMDS. The header may also contain information to be used for congestion management and for error detection for control messages.

  Unlike frame structures, multiple framing bits or characters are *not* used to provide fill between cells when there are no data to transmit. Instead, entire empty (or null) cells are transmitted.
- **Payload**, the "information" portion of the cell. The most fundamental characteristic of a cell is the fixed size of the payload. In the case of ATM, the size of the payload is 48 octets (8-bit bytes). The payload consists of the traffic being transported in its native format. However, since the native format will seldom be the exact same size as the cell, it usually must be segmented into several cells.

A frame is composed of:

- The frame **flag**, used to separate one message unit from another. In frame-oriented technologies, the flag will usually be an eight-bit character with the hexadecimal value of 7E. There is generally a minimum of one flag between frames. Multiple flag characters are used to provide fill between frames when there are no data to transmit.

- ✔ The **header** portion of the message unit; this provides the identification of the owner of the information. In frame relay and ATM, the header is used to identify the circuit number on the individual link. The switches then provide connections among "circuit numbers" to route message units to their appropriate destinations.
- ✔ The **payload**, the "information" portion of the frame. The most fundamental characteristic of a frame is the ability of the payload to be variable in size. The minimum payload size is typically a single character. There is a maximum size, and this is often negotiated at call setup. The typical maximum sizes range from about 2,000 to about 8,000 characters.

  The payload consists of the traffic being transported in its native format. A single payload may contain an entire native-mode message unit if the message unit will fit within the maximum size constraints. If the native-mode payload is larger than the maximum size for the payload, it must be segmented into several frames.
- ✔ The **CRC** or critical task of error control; in a broadband packet system this ensures the integrity of control messages and that the header was not corrupted. This may be accomplished by placing error control either in the header itself or in the trailer.

Once again, remember that these characteristics are general, not always true. Neither frames nor cells are inherently better for all applications. Even for an application that is generally better served by one of the technologies, exact performance will be affected severely by the actual implementation.

Interface Standards

An oft-recurring concern with all of the broadband packet technologies is whether there are any standards. There are full suites of standards in place and/or in definition for each of the technologies. None is being developed as proprietary or vendor-specific.

Most of the standards in place are User-to-Network Interface (UNI) standards. As such, they define the rules for passing information from a piece of data terminal equipment (DTE), such as a bridge, router, broadband packet "pad", broadband-packet concentrator, or front-end processor, to a packet-transporting network. This network may be a carrier network or a private network owned and operated by the user. Its function is to transport broadband packet of some type from the entry point to the destination by whatever means it deems appropriate.

The internal structure of the network, often called the "network architecture," is generally not subject to standards. Each network may transport information among the network nodes in any manner the network designers choose. In fact, most networks will use a proprietary internodal transport mechanism to provide a "better mousetrap" to the industry.

This proprietary internodal transport may appear a severe limitation, but is no different from what occurs in most other standards. It is exactly like X.25, also a UNI specification. Each packet network uses it own internodal transport mechanisms, as shown in Figure 6.14.

FIGURE 6.14 Broadband transport network standard interfaces

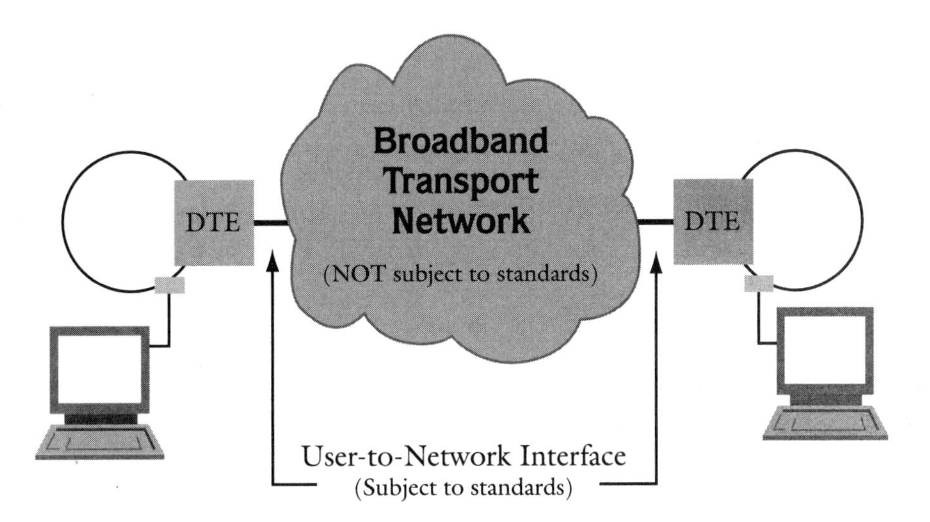

If the entire network architecture were standardized, there would be little or no incentive for equipment manufacturers to build products that are better, and not just cheaper. Thus, the proprietary nature of the network architectures allows competitive factors in the market to continue to advance the technology.

Data transfers between disparate X.25 networks are accomplished via X.75, the Network-to-Network Interface (NNI) specification. Similarly, NNIs developed and/or are being developed for each of the broadband packet technologies.

The standards work is not over, though, and there are several areas in which it continues. The most important point to remember is that many broadband packet standards are here, are real, and are functional. Standards are living documents, though, so they will continue to evolve.

Broadband Packet and the OSI Model

We must also realize that the broadband packet standards are not designed to accomplish all of the tasks in the network. In reality, they address only a very narrow subset of the OSI seven-layer protocol stack (see Figure 6.15).

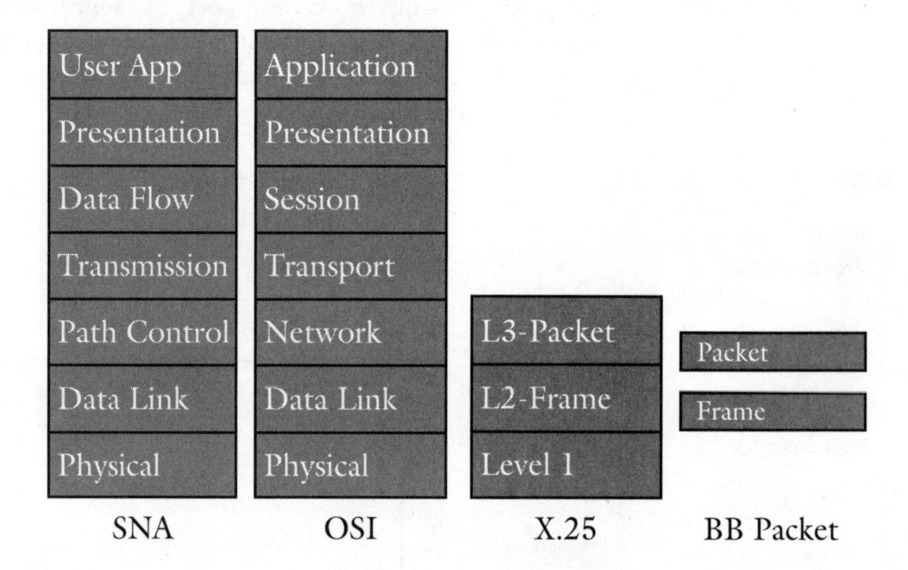

FIGURE 6.15 Positioning of broadband packet

Broadband packet technologies may be best thought of as the WAN equivalent of a media access control (MAC)-layer protocol in the LAN world. The standards define how multiple applications may share the transmission bandwidth in a common, packetized format. Conformance to any particular higher-layer protocols and protocol translation is not part of broadband packet's task set.

As Figure 6.15 indicates, broadband packet accomplishes some of the traditional Level 2 and Level 3 tasks. These are all that is needed to fulfill the mission. While some of the technologies are designed for, or at least recommend, a particular physical level implementation, the fundamental transport tasks for broadband packet can generally be accomplished with a variety of physical-level implementations.

What's Next

This chapter discussed broadband packet networks and voice communication. The next chapter treats the digitization of voice, both the technology and the standard, as well as implementation issues.

PART II

Hands-On VoIP: Standards and Implementations

CHAPTER 7

Codecs Methods

This chapter provides an in-depth discussion of codecs technology as well as of vocoding methods.

Audio/Video Codecs Review

The basic functionality of a video codec (COder/DECoder) is to enable the transmission of various audio, video and data signals over digital telephone networks. In general terms, an audio/video codec embodies the functionality depicted in Figure 7.1

FIGURE 7.1 Audio/video codec functionality

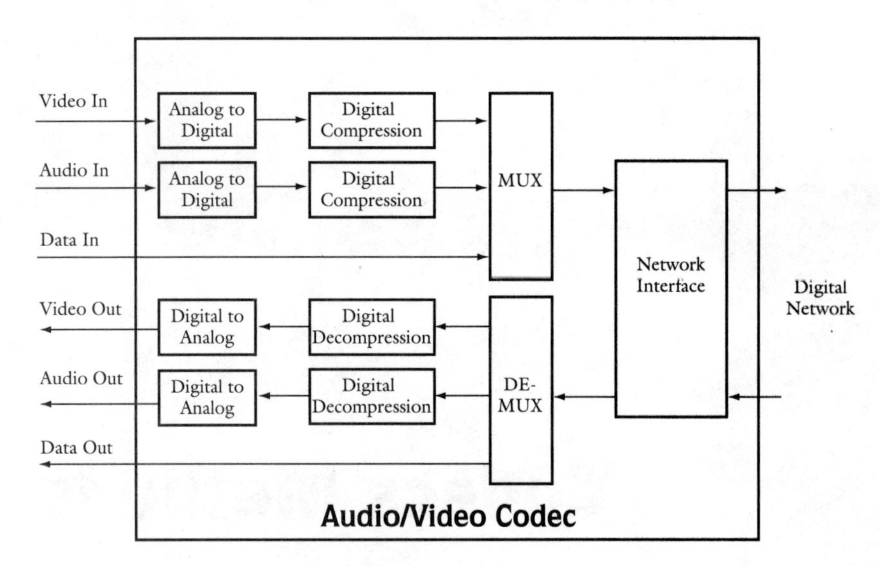

A codec is comprised of two basic processing elements: a compressor (or encoder) and a decompressor (or decoder).

On the compressor side a standard video signal coming from cameras or tape players is first digitized into a 135 Mbps feed. Since this rather high bandwidth is difficult to transport economically, it must first be compressed to a more manageable bit rate. While many compression techniques exist, most codecs on the market rely on either Delta Pulse Code Modulation (DPCM) or Discrete-Cosine-Transform (DCT) algorithms. Some compressors achieve very high compression ratios, up to 200:1. Others compress as little as 5:1, but maintain very high picture quality.

The same process is applied to audio signals, which are first digitized, and compressed. The compressed digital video data are then multiplexed with the compressed audio data and, in some cases, external digital data. The resulting data stream is then formatted according to a given network interface standard, and connected to the network.

The decompressor side reverses the process. It demultiplexes the original, compressed digital video, audio and data streams, and feeds the resulting signals to the respective decompressors. The decompressed signals are then converted back to their original, analog form.

Audio Codecs

An audio codec is a software scheme that takes analog audio data and encodes it into some sort of binary format for digital storage and/or processing. It then decodes the data and attempts to reproduce the original sound. Due to the massive size of digital audio data, codecs usually involve some sort of compression. Audio codecs can be broadly divided into three classes: waveform, source and, hybrid.

Waveform Codecs

Waveform codecs try to save enough data about the original sound wave to enable it to be reconstructed upon playback, by reproducing the sound wave itself. This has several advantages over the other two types of audio codecs, as they try to synthesize sound.

Another advantage of waveform codecs is that they are device-independent, which in theory enables them to reproduce all sorts of sounds, regardless of source. High-end computers are not required to run this type of codec, as its software to encode and decode is not as complex as the others. A 486 Intel box can be sufficient. Of course, with the price of computers decreasing, a Pentium-based computer might be useful, as the audio codecs can be run quickly and easily, with almost instantaneous playback.

The only disadvantage to waveform codecs is that the files tend to be large. Windows 95 and NT offer several types of waveform codec, including PCM, companding and ADPCM codecs.

Defining Pulse Code Modulation (PCM)

PCM is the simplest form of audio codec. It is also the method used by Microsoft Windows when automatically saving a WAV file. PCM files do not use any form of compression. A music clip digitalized at 22,050 Hz 8-bit mono (radio quality) will require more than 22 KB of file size per second of audio when saved in this format. CD-quality sound files in this format are eight times larger. However, PCM-format WAV files are natively supported by virtually every sound card on every platform.

To learn more about the method in which the PCM codec digitalizes audio, visit the SK Web construction site at **http://www.skwc.com/WebClass/SoundProcessing.html#pcm** where there is a comprehensive description of the subject.

Defining Compacting/Expanding (Companding) Codecs

Companding codecs, based upon the A-Law and μ-Law standards, were developed to address accurate reproduction of sounds for humans. Companding codecs have been around since the 1960s and are still in widespread use today. They are not really compression codecs. Although they are similar to PCM, also not compressed, the main difference is that the companded file will have an apparent amplitude range not far below that of a 16-bit file. Thus, companding codecs are most useful for reducing 16-bit audio files to 8-bit ones, cutting their size in half without sacrificing as much quality. Windows 95 and NT offer both an A-Law and μ-Law version of these codecs.

Defining Adaptive Differential PCM (ADPCM)

A waveform produced by human speech. is easy to predict in the short term. One of the most common techniques in coding speech samples uses this fact and attempts to predict the value of the next sample from the values of the previous samples. If the predictions are accurate, charting the difference between the predictions and the actual samples will produce a much flatter graph than charting the wave itself.

This technique of charting differences is known as Differential Pulse Code Modulation (DPCM). The differences can be made even smaller if the predictor can be made adaptive, so that it will change its predictions to match individual characteristics of the speech being coded. This is the concept behind Adaptive Differential PCM codecs. But since these codecs are built to predict the patterns of human speech, they do not give very high quality music reproduction. Also, this is a very complex process, especially if compared to a sound wave. Decoding and (especially) encoding requires more work by a computer's CPU and usually results in delay.

Source Codecs

Source codecs attempt to create a model of how a sound was generated and then try to reconstruct it based upon that model, discarding the waveform data completely. Vocoders are the most typical source codecs, probably the only common ones, as they are constructed on a basic model of how the human voice is produced. They save a few parameters (guidelines) based upon the individual characteristics of the particular voice, which they will attempt to reproduce. Since vocoders save a very small set of characteristic traits and completely discard the waveform data, they produce very small file sizes.

The limitation of source codecs is that they can only save human speech data, and only one voice speaking at a time; the synthesized output tends to be artificial, especially if compared to other methods. They are not recommended for reproducing any other sounds.

AT&T's Bell Laboratories has a site where a word can be typed into an online form and a button clicked; the word is then spoken by an artificial voice as a man, a woman, a child, or a gnat. To try it out, visit **http://portal.research.bell-labs.com/cgi-bin/ voices.form**.

Hybrid Codecs

Hybrid codecs usually produce smaller files than waveform codecs. However, they are more complex and sound reproduction quality is not as good. Still, hybrid codecs are less complex and produce better quality sound than vocoders. They are called hybrids because they utilize elements of each of the other two approaches.

The most successful hybrid codecs are the Analysis-by-Synthesis (AbS) codecs. Like vocoders, they begin with a synthetic model of the human vocal tract—vocal chords, throat, mouth, teeth, tongue, lips...—and attempt to reproduce the human voice. AbS codecs differ from vocoders in that they use data from the actual sound waves to select an excitation signal from a number of built-in options.

Installing Audio Codecs on Windows 95/NT 4.0

If you are running Windows 95, be sure to have the Microsoft Audio Compression Manager installed. Go to the control panel and double-click on the multimedia icon. Then select the advanced tab, as shown in Figure 7.2.

Once in the advanced tab, chose Audio Compression Codecs and click on the plus mark (+) to its left. This should open up to reveal a number of options, as shown in Figure 7.2. If all the options shown in Figure 7.2 are not available they can be added by clicking over the icon Add/Remove Programs, as shown in Figure 7.3, and then clicking on Windows setup, which should produce a tab just like the one in Figure 7.4. Once there, select the Audio Compression option, as illustrated in Figure 7.5.

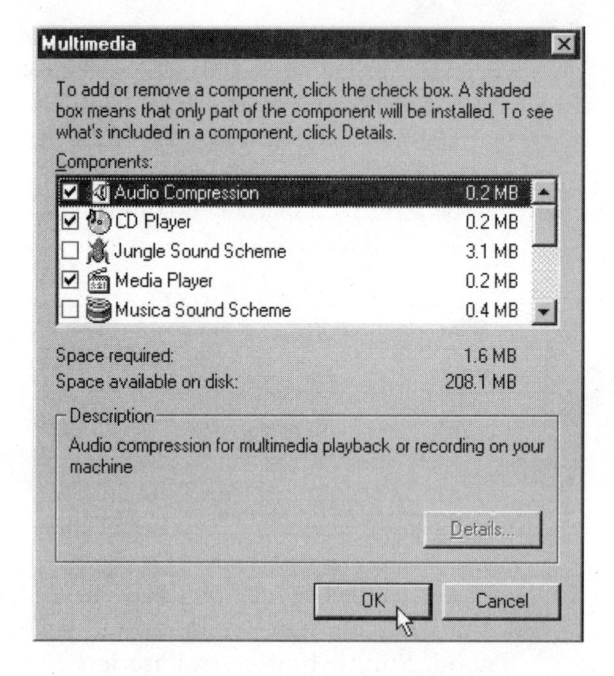

FIGURE 7.2 Installing audio codecs on Windows 95/NT 4.0

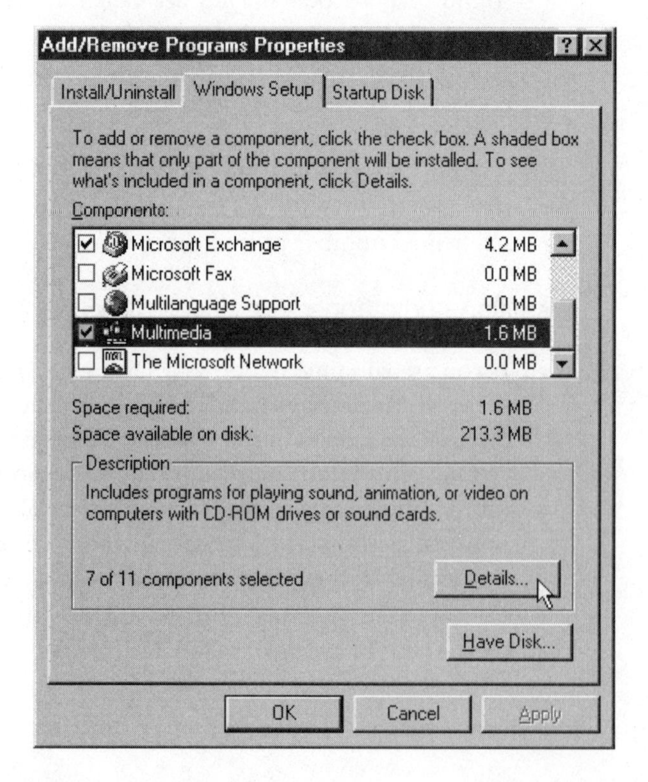

FIGURE 7.3 Adding audio codecs options on Windows 95/NT 4.0

FIGURE 7.4 Selecting additional audio codes in multimedia setup on Windows 95/NT 4.0

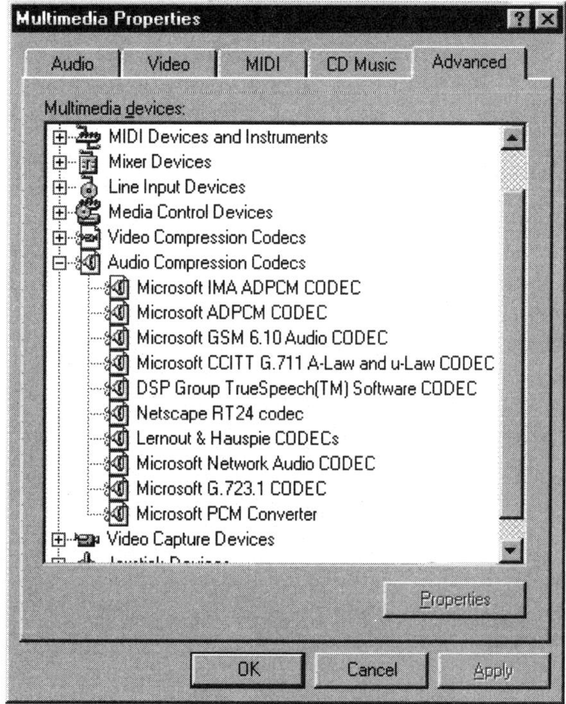

FIGURE 7.5 Choosing audio compression options on Windows 95/NT 4.0

What's Next

In the next chapter the focus will be on the applicability of VoIP, in computer telephony integration (CTI), videoconferencing, document-sharing, web-based call center applications, etc. The challenges VoIP faces, getting telcos up to speed, and setting standards are treated. The major VoIP players, including 3Com, Motorola, Nuera Communications and others are discussed.

CHAPTER 8

Voice Over IP: Can We Talk?

VoIP is transforming the trillion-dollar global communications industry. In a few more years at least 70 percent of all voice traffic will become packet-based, traveling over the Internet. The movie *Deep Impact* in early 1998 gave an idea of what will happen to many of the existing telcos sharing the voice traffic. My assessment is that the tides of VoIP are becoming so high (so attractive) that it will sweep away many of the telcos. Until they can reengineer their way of business and the way they carry voice, to become what the VoIP Investor's Page (**http://home1.gte.net/denunzio/**) call "NextGen Communications Companies," these companies will experience many changes and after-impact effects.

Let us consider VoIP applications and what some of the main players are up to.

VoIP Applicability

Most implementations of VoIP focus on the first two benefit areas: cost-savings/simplification, and extending the corporate telephone infrastructure to small sites. Computer-telephony applications will certainly emerge, but adoption will be slower because these applications require changes in workplace processes and possibly in computing infrastructure.

Applications using Motorola's Vanguard Voice (other vendor applications exist, as we will see) are found in the following:

- ✔ **Small Office/Home Office Connectivity:** Small branch offices and home-based employees are taking advantage of inexpensive ISP connections and VoIP technology to gain transparent access to large-site computing and telephone functions
- ✔ **Intranet Telephony:** A large nationwide bank in Latin America is using VoIP to carry telephone calls between headquarters and some 800 branch offices
- ✔ **DTMF Tone-Based Security:** Many regionally-dispersed companies (in this case a savings bank) use DTMF tones ("touch-tones") as a way for security monitoring personnel to enter status codes as they patrol from site to site. Traditionally, these systems require expensive leased lines, or they generate numerous PSTN calls. With VoIP, the bank now routes the code-entry over the IP network that already reaches all the necessary locations.
- ✔ **Next-Generation Telecommunication Service Providers:** The advantages of VoIP have not been lost on telecommunications carriers. One example is Net Communications Inc., a "next-generation" telco providing international voice services over its high-bandwidth IP backbone. Another is a satellite network operator providing IP telephony connections between Latin American companies and their U.S. sales offices.

Other areas of applicability for VoIP include computer telephony integration, videoconferencing, document sharing and more.

Computer Telephony Integration (CTI)

One of the best examples of CTI is SoundWare's Telephony Operating System (TOS), the original telephony on audio software platform and one of the leaders in the industry. It enables high-level PC telephony to be implemented without costly, complex hardware and software. TOS is a Telephony Application Program Interface (TAPI)-compliant Windows extension for managing host-based signal processing/communication algorithms (hardware replacements) running on x86 and Pentium Pro processors (MMX). TOS also supports all major audio chip sets and is expected to play an important role in the migration to digital audio/processing solutions.

TOS enables telephony applications to be migrated to the SoundWare TOS environment. Developers can continue using standard Windows APIs. TOS supports MMX based SPMs for computation-intensive algorithms such as modem data pump, digital mixing, and on-the-fly sample-rate conversions. TOS manages all aspects of call processing, providing a simple, flexible, and open telephony platform for developers of signal processing and communication algorithms.

But SoundWare is not alone; many other suppliers offer similar products and services. Some are featured in this chapter and Chapter 9. For now, let us continue to assess the applicability of VoIP.

Videoconferencing

Video conferencing is becoming more available and affordable every day. It is relatively easy to use and of high quality. However there are some factors to be aware of. Video conferencing is about bandwidth, or how much information can get through the pipeline, how much information can be compressed, pushed through the pipeline, decompressed, and how fast. Below are ways that Vivid Communications (**http://www.vividcommunications.com/Web_store/web_store.cgi**) describes as means of pushing videoconferencing to recipients:

- ✔ **POTS (Plain Old Telephone System)**—Great for face-to-face; cannot fit data and video at the same time
- ✔ **ISDN-BRI (Basic Rate) up to 128 Kbps**—Two 64Kbs-lines that can be banded together; best for the small-business owner and telecommuters—decent frame rates and data sharing capabilities
- ✔ **ISDN-PRI (Primary Rate) up to 356 Kbps**—More expensive, but if the best picture is the object, this is the way to go
- ✔ **Internet**—Always depends on how many people are on the Web and how good a connection is
- ✔ **LAN / WAN**—New multiplexers help push data through so as not to slow down LAN systems. Excellent frame rates can be accomplished over LAN systems.

The smaller the bandwidth of the pipeline, the more compression/decompression and speed will be needed. Some systems accomplish this with software and some with software and hardware. Computers with a Pentium chip and at least a 28,800 modem are the best bets. The faster the Pentium chip the faster compression/decompression can be accomplished. Systems that take advantage of MMX technology will improve the frame rate as well.

Video capture cards usually work as buffers. They will watch the picture and work only with parts of the picture that have changed. In this way the computer does not have constantly to reread an entire picture, but only reads the parts that have changed. This speeds up the capability of the system to read the parts of the picture that *have* changed. There is less information to be decompressed and the computer is not working with redundant parts of the picture.

The size of the picture will affect frame rate as well. The smaller the picture and the fewer pixels the system will have to read, the better the picture. Another factor is the camera. CCD cameras are digital cameras developed to deliver voice with picture (lip sync). They require a video capture card. Some less-expensive digital cameras plug into your parallel port and are fine for Internet—family fun, but are less precise about getting the voice and the picture together.

Currently, the best business video conferencing if you don't have a LAN system is through ISDN lines. Data sharing and whiteboarding (plain whiteboard that can be written on by both ends) are possible over the ISDN bandwidth, while video conferencing proceeds. More bandwidth is available by binding together two or more ISDN lines, though some products do not have this ability.

Standards have been set by the ITU for the video conferencing industry that allow different product brands to work together. Although the standards have been set some systems are still proprietary (the same software is needed on both ends to communicate). The standards-based systems are POTS H.324, ISDN H.320, LAN and Internet H.323. Standards have been set for voice and data sharing as well.

Turning a laptop computer or a desktop computer into a complete boardroom setup can be easy and economical. Try displaying the conference over one of the Proxima or In Focus overhead projectors, which can be used for presentations of all kinds.

TIP

Vivid Communications has a comprehensive book on video conferencing, by Evan Rosen, Personal Videoconferencing. It is very user friendly and gives an in-depth look at the current trends and the future of video conferencing.

PicturePhone is another leading vendor supplying videoconferencing products and services, providing: ISDN, LAN/WAN and Internet, Analog, and Switched 56, 384 and T1 video conferencing.

TIP

For more information on PicturePhone products and solutions, visit their Web site at **http://www.picturephone.com**.

Document Sharing

Document sharing, through technologies such as the compound document framework, provides support for synchronous collaboration between multiple users of a document. Several users can edit a document in real time—each seeing the other's changes as they occur.

Netopia's Virtual Office is a good example of document sharing technology. The product allows you to collaborate on a document in real time, while remotely connected to the Internet via ISDN routers and LAN/WAN office networks. According to Netopia, NVO works when there is a decent Internet connection. They recommend at least a 56K-bit/sec. connection and a recent browser, able to run Java applets, because NVO relies on these for its controls.

TIP

For more information on Netopia's Virtual Office go to Geocities (**www.geocities.com**) and set up a free Netopia account for the first year.

Web-Based Call Center Applications

Web-based call centers are a fast-growing VoIP-derived application. Micron Electronics implemented their Web-based call center in Spring 1998, and they are not alone. The convergence of call centers and the Internet is taking place in the space defined by the use of VoIP technologies, interactive Web browsing, and the integration of existing call-center technologies such as automated call distribution modules and PBX switches with IP gateways.

Some early products only use the Web to initiate a callback by an agent, but the majority of the latest Internet call-center solutions are integrating the Web with the back office of a call center, where the Internet becomes truly an extension of a call center.

Lucent Technologies, NetSpeak and eFusion are companies offering Web-based call-center products and solutions, with starting prices at

$100,000. Most of the applications work along this paradigm: a customer navigating through sales or technical support information wants more detail. A button on the page connects this customer to an agent via VoIP or text-based chat. In the most fully integrated Web-based call center, user information is retrieved at the time of the call; this can come from v-cards, cookies, log-on information, or forms.

But IP technologies are not going to overrun existing call centers. I believe the existing PBXs are going to stick around and VoIP will just complement them, providing a powerful combination of IP and PBX in the same server environment.

VoIP Challenges

Not surprisingly, VoIP has some challenges ahead. From getting telco investment through setting and agreeing on a standard, it will be some time before we see major developments in this area. Let us consider some of these challenges.

Getting Telcos Up to Speed

During Fall 1997, Qwest Communications, a Denver telco, began offering long distance phone services to consumers at 7.5 cents-per-minute. The most surprising news to the telecommunications industry was not the fact that Qwest had undercut the competition by 50 percent, but that the company was using VoIP technology.

Qwest's move got telcos talking. AT&T, Sprint and even WorldCom argued that VoIP was not ready for prime time. As of Spring 1998, no one else is yet using VoIP, so the company is gaining a lot of momentum with this new technology. Qwest is planning to offer not only long distance phone services, but also virtual private networks and concurrent engineering, where engineers collaborate over the network using high-bandwidth CAD images, especially since bandwidth is not a problem.

Telcos will have to get up to speed, not only with the technology, but also by convincing their boards of directors and stockholders that VoIP is here to stay. Meanwhile, some of them are trying to stall Qwest (and the technology!) through lawsuits.

Setting Standards

What follows outlines the main protocol standards used with VoIP technologies, instrumental in the development of open telecommunications.

H.323

The International Telecommunications Union (ITU) recommendation H.323, issued May 28, 1996, describes how terminals and equipment can carry any combination of real-time voice, data, and video, including video telephony, over a local area network.

H.323 is sometimes referred to as an "umbrella" recommendation, since it contains references to most of the other recommendations, including:

- ✔ H.225.0 packet and synchronization
- ✔ H.245 control
- ✔ H.261
- ✔ H.263 video codecs
- ✔ G.711, G.722, G.728, G.729, and G.723 vocoders
- ✔ T.120 series of multimedia communications protocols.

Together, these specifications define a number of new network components (H.323 terminal, H.323 Multipoint Control Unit (MCU), H.323 gatekeeper and H.323 gateway), all of which interoperate with other standards-compliant endpoints and networks by virtue of an H.323 gateway.

The H.323 specification includes a list of vocoders allowed in H.323-compliant clients and gateways. The G.723.1 vocoder has been specified as the default vocoder for H.323; all clients and gateways must support G.723.1. This guarantees interoperability at the vocoder level. Once two entities have established that they both support G.723.1 during the call setup, they can negotiate to find a mutually preferred vocoder and can use that instead.

Virtually every company in the IP telephony market has announced plans to be H.323-compliant. This list of companies includes Microsoft, Netscape, and Intel, providers of the most widely deployed client software in the market. Natural MicroSystems supports H.323 in the Fusion IP telephony development platform (see Chapter 9 for more information on Natural MicroSystems.)

H.100/H.110

H.100, approved on May 23, 1997 by the Enterprise Computer Telephony Forum (ECTF), is a standard designed to spur the growth of the CT.

H.100 provides for a single telecom bus superseding all existing bus architectures including MVIP and SCbus.

The H.100 bus is interoperable with the current industry telecom buses, allowing developers to integrate newer H.100-based products with existing products. It is the telecom bus standard to use in conjunction with the PCI bus standard for personal computers but easily interconnects with telecom buses on ISA/EISA boards.

H.100 supports an 8-Mbps data rate and 128 channels per stream for greater bandwidth than that provided by previous telecom buses. H.100 provides a total of 4,096 bi-directional 64-kbps timeslots, permitting up to 2,048 full-duplex calls. This compares to 512 timeslots for MVIP-90, 1,024 for the SCbus in PCs, 2,048 for SCbus on VME, and 3,072 for H-MVIP.

The increased number of timeslots provides greater communications capacity, due to the introduction of a 68-pin fine-pitch ribbon cable which is physically smaller than the existing 40-pin regular-pitch MVIP cable. H.100-based boards can be interconnected with MVIP or SCbus boards via a passive transition device, commonly called a “swizzle stick,” that allows the connection of the different ribbon cables.

In systems comprised of a combination of boards, the master clock must be an H.100 board. H.100 master-clock circuits also include compatibility clocks for driving existing MVIP and SCbus boards that operate in clock/slave mode. To facilitate operation with MVIP and SCbus boards, H.100 allows individual data lines to be programmed in groups of four to operate at 2, 4 or 8 Mbps, allowing direct connection to existing boards at their native operating speeds.

The H.100 specification incorporates technology from GO-MVIP (Global Organization for MVIP) such as the programmable operating speeds technique of H-MVIP and redundant clocks from MC1 multi-chassis MVIP. As mentioned, programmable operating speeds provide support for the interoperation with MVIP and SCbus boards. The redundant clock eliminates a single point of failure. If any telecom board fails, including the H.100 master clock, the system will continue to operate.

H.100 offers developers and integrators extensive new capabilities. It brings more capacity than any existing bus, enabling developers to deliver larger and lower-cost applications.

H.110 is the CompactPCI version of the H.100 standard. This standard allows hot-swapping of boards in CompactPCI chassis from the PCI Industrial Computer Manufacturer’s Group (PICMG), offering customers the option of a CT system with virtually no down time. This enables automated call centers, IP gateways, and voice messaging systems to remain up and running until there is a convenient time to replace a failed board.

MVIP

Multi-Vendor Integration Protocol (MVIP) is the de facto industry standard hardware and software architecture for platform interoperability and telephone switching among ISA and PCI-based computer systems. Natural MicroSystems and six other companies developed the original MVIP standard in 1990. The most widely deployed standard for interoperability among computer telephony vendors, MVIP is now maintained by GO-MVIP, an independent organization.

The MVIP family of standards addresses both configurations of networked PCs and single computer chassis configurations. Several hundred companies support MVIP's open scalable, switching architecture. Over 200 MVIP-compatible products are available on the market.

CompactPCI

CompactPCI, initiated in 1994 by Ziatech Corporation under the auspices of the PCI Industrial Computer Manufacturers Group (PICMG), is the newest specification for PCI-based industrial computers and defines many features that make a PC more available.

CompactPCI offers a host of telecom features required for network applications including:

- ✔ A standard telecom bus (32 streams and 4096 time slots) for communications between cards in a chassis rack
- ✔ A telecom form factor (6U card heights with rear panel I/O)
- ✔ Hot-swap capability with staged pins and system notifications with card-tab release, allowing systems to be upgraded or expanded, or cards replaced without taking servers off-line
- ✔ No interruption of system operation if a subsystem module fails
- ✔ Redundant chassis and board configurations for highly available resource requirements
- ✔ Redundant power management, CPUs, and disks
- ✔ Software compatible with mass-market PCI systems
- ✔ Telecom power bus (48 V DC) and provision for ringing voltage
- ✔ Transition cards and cabling assemblies to simplify installation.

The principal benefits of CompactPCI include:

- ✔ An open, industry-accepted specification eliminating the proprietary nature of previous high-availability systems
- ✔ Compatibility of software from standard PC (PCI) systems to CompactPCI systems—existing PC software can run unchanged on CompactPCI systems

- ✔ Delivery of hot-swappable telecom features required for
- ✔ Higher reliability through hot swapping of components.

What's Next

The next chapter assesses what is being offered by VoIP innovators such as NetSpeak, NetPhone, Vocaltec, TeleVideo Conversions, Inc., Vienna Systems, Lucent Technologies, among others.

CHAPTER 9

What to Expect: The Innovators

This chapter provides a brief profile of the major players in VoIP technology, and a technical overview of the main VoIP products available on the market as of Spring 1998. I included an extensive selection of all the major players and their products to allow readers the chance to evaluate each before deciding which product best suits their needs.

This selection includes many different VoIP technologies and products, from 3Com's Total Control Hyper Access system and Motorola's VIPR, to Nuera Communications' Access Plus lines and a series of Internet phones.

I cannot recommend any particular product since the requirements and features of VoIP products change depending on the environment. My preferences are probably biased, because they are directly related to the environment I work with. All the information in this section was provided by the vendors mentioned. Some provided more information than others. Do not opt for any of these products based on the number of pages or details provided here. Most of the vendors listed also provided demo and/or evaluation copies of their products in the CD that accompanies this book. Refer to it for additional information.

Some VoIP Major Players

An informated decision when selecting a VoIP product or service to best suits particular needs, will be aided by careful reading of this chapter, and a tabular summary of all the features needed. Then, check the CD and the products selected and run a complete "dry-run" on them before making a decision. Do not forget to contact the vendor directly, as these products are always being upgraded and new features incorporated, and that could make a difference in any decision. Contact information and a brief background about the vendor are provided at the beginning of every section.

3Com's Total Control HiPer Access System

3Com provides a single platform that combines multiple services for real-time interactive communications to VoIP using standard telephones all the way to VPN, multimedia and more. The Total Control multi-service access system leverages the inherent strengths of 3Com's Total Control platform's HiPer DSP technology and EdgeServer Pro module to offer a powerful integration of Internet and voice technologies.

3Com's Total Control System: Maximizing Internet Technologies.

3Com's Total Control System provides simultaneous voice and Internet access over the same connection for the ultimate in collaborative computing, customer support, increased productivity and cost savings.

One of the main strengths of the Total Control chassis is the ability to add custom features to the powerful DSPs on the HiPer DSP card. Each HiPer DSP card can handle 24 phone lines (30 in Europe), and the EdgeServer VoIP system can handle 13 HiPer DSP cards. On each DSP card, the 12 Texas Instrument TMS320C548 DSPs process data at 100 MHz each.

These large computer engines allow for preprocessing of most of the bit-intensive work on the DSP. The preprocessing allows the Total Control chassis to forward packets quickly and to handle large numbers of ports since the work for each channel is handled in parallel. When all CPUs involved in a fully configured EdgeServer Total Control chassis are lumped together, nearly 34 billion instructions per second (34,000 MIPS) are used.

This massively parallel approach has been used in the past by 3Com to allow the Total Control remote access servers to pre-process PPP packets to achieve high packet rates. These DSPs can be programmed on a per-call basis to handle an incoming call as a modem or ISDN call for remote access, a fax, VoIP or video conferencing call. This flexibility to add new functionality to an open programmable DSP provides investment protection for the user. It also enables system operators to provide multiple revenue-generating services on a common access platform. Because VoIP and Video use the existing HiPer DSP cards that are used in the Total Control Remote Access Concentrator, the incremental cost to add these features can be kept low.

3Com shines with its carrier-class VoIP/video solution, the EdgeServer (**www.edgeserver.com**). The EdgeServer is a complete PC server that runs open operating systems and fits in the Total Control chassis. Currently only Windows NT 4.0 supports full connectivity to the HiPer DSP cards on the EdgeServer.

Windows NT 4.0 provides a full-featured and stable platform with the standard telephony interface TAPI. TAPI is a connection layer that allows telephony applications like call control to interact with telephony-enabled devices like the HiPer DSP drivers. TAPI version 3.0 enables very large telephone systems to be built via master/slave TAPI. The slave TAPI process would run on the EdgeServer driving the HiPer DSP cards in each Total Control chassis. The master TAPI process could reside on a local standalone PC and act as the overall call control agent for the entire slave TAPI chassis. The master TAPI agent could also be in a remote location providing a single point of control for geographically dispersed chassis.

In addition to providing the industry-standard TAPI interface for customer customization, EdgeServer has a wide variety of useful development tools such as Visual C++, Visual Basic and Borland's Delphi.

For more information on 3Com's VoIP Products, visit **http://www.3com.com**.

3Com and eFusion: Enhancing VoIP

Enhanced VoIP features, such as call forwarding, call waiting and other business-oriented capabilities are becoming reality already. As I write this section 3Com is engaging in a partnership with eFusion Inc. to develop a series of VoIP applications based on Total Control remote access concentrator.

Although this book will be released before the scheduled release of the product, in the fourth quarter of 1998, make sure to seek more information about it, as 3Com plans to support a variety of eFusion applications, including those tailored to enhanced Internet telephony services, as well as e-commerce applications such as its Push-to-Talk voice and Internet browsing product. All the applications are still H.323-compatible, ensuring operability across product lines.

Sound Design's SoundWare

The Telephony Operating System (TOS), is a patented TAPI-compliant Windows extension for managing host-based signal processing/communication algorithms (hardware replacements) running on x86 and Pentium Pro processors (MMX). TOS supports all major audio chip sets and is expected to play an important role in the migration to digital audio/processing solutions.

Telephony applications can be migrated easily to the SoundWare TOS environment. Developers can continue using standard Windows APIs. TOS supports MMX based SPMs for computation intensive algorithms such as modem data pump, digital mixing, and on-the-fly sample rate conversions. TOS manages all aspects of call processing, providing a simple, flexible, and open telephony platform for developers of signal processing and communication algorithms.

SoundWare runs on any standard PC sound subsystem without requiring a modem, enabling a SoundWare "phone ready" multimedia PC or sound card to add telephony and communication functions as mainstream features. SoundWare provides PC bus independence, scalability via host-based algorithms, full hardware and software integration with the standard PC audio channel, and complete software integration with the Windows multimedia, telephony, and communication APIs. In order to make

SoundWare-ready offerings, PC OEMs and audio vendors are adding a simple, low-cost universal audio link header to their existing mother boards, sound cards, and combo cards without making any architectural changes.

For more information on Sound Designs Soundware product, visit **http://www.soundesigns.com**.

Natural MicroSystems' Fusion

Natural MicroSystems (**http://www.nmss.com/nmss/nmsweb.nsf/nmshomeview/nmshome**) is one of the leading suppliers of open telecommunications-enabling technologies for developers of high value telecommunications systems and applications built on standard computing platforms.

Open telecommunications are concerned with moving the telecommunications equipment business to open, mass-market computing platforms. Natural MicroSystems builds hardware and software component products that enable developers of telecommunications equipment to build their applications and systems using these open platforms. Their solutions are based on open standards that make it possible for their partners to quickly develop high-performance, high-capacity multimedia communications systems and applications. Natural MicroSystems helps its partners create new markets, new value for their customers, and new opportunities to grow.

The value of open telecommunications is the ability it gives developers to take advantage of standards-based building blocks, reducing time-to-market when creating complex applications. The Natural MicroSystems family of products provides fundamental functions that adhere to widely accepted standards for easy integration into communications networks around the world. Their solutions are based on standard computing platforms, such as PCs, so that developers can take full advantage of the wealth of products, tools, and support that open systems provide while also offering truly global communications products.

For more information about Natural MicroSystems, visit their Web site at **http://www.nmss.com/nmss/Nmsweb.nsf?OpenDatabase**, or contact them at: 100 Crossing Blvd., Framingham, MA 01702-5406; Tel: 800-533-6120 or 508-620-9300; Fax: 508-620-9313.

Fusion

Fusion is the industry's most scalable, highest-performance PC development platform for standards-based IP Telephony gateways. Fusion is compliant with both the International Telecommunications Union's (ITU's) H.323 specification and the International Multimedia Telecommunications Consortium's (IMTC's) VoIP Implementation Agreement. Fusion enables developers to create gateways with configurations from 8 ports to multiple T1s/E1s with no increase in latency or decrease in performance. Building on its basic configuration of a full T1 of IP Telephony in only two ISA slots, Fusion's scalable architecture supports the highest port capacity of any solution on the market.

Fusion uses an intelligent hardware and software architecture that integrates Public Switched Telephone Network (PSTN) interfaces, telephony protocols, speech encoding, LAN interfaces and data protocols into a cohesive, flexible package. Incorporating a series of highly accessible Application Programming Interfaces (APIs) and a group of dedicated boards that communicate with each other via the industry-standard Multi-Vendor Integration Protocol (MVIP), Fusion offloads key processes from the host CPU and memory.

Features

Fusion's main features include:

- ✔ It supports the highest capacity (multiple T1s/E1s) of any standards-based H.323 development platform
- ✔ It offloads processing from host CPU and memory to minimize latency, maximize scalability, and free resources to run higher level applications
- ✔ It supports ITU's H.323 specification and IMTC's VoIP recommendation, enabling interoperability with other H.323-compliant clients and gateways
- ✔ It supports gatekeeper functions for address translation, control access and bandwidth management
- ✔ Its high performance/low latency does not degrade as a system scales from 8 ports to multiple T1s/E1s
- ✔ It supports the broadest choice of standard vocoding algorithms, including G.723.1, G.729A, MS GSM, and VoxWare MetaVoice RT24
- ✔ It embeds standard Internet protocols on a dedicated board for maximum performance, including TCP/IP, UDP and, RTP/RTCP.
- ✔ Its compact footprint supports T1/E1 of IP telephony capability in only 2 ISA slots

- ✔ It utilizes MVIP, an industry-standard, dedicated TDM bus that acts as a switching fabric among the DSP resources, PSTN interfaces, and LAN ports, easing integration with other MVIP-compliant products
- ✔ It supports a broad range of industry-standard clients, including Microsoft NetMeeting, NetScape Conference, VoxWare VoxPhone, Netspeak WebPhone, and VocalTec
- ✔ It is ideal for toll bypass, voice and fax messaging, LAN telephony, Web-enabled call centers, interactive voice response (IVR), and remote teleworking applications.

Fusion integrates hardware and software within a standard PC, greatly simplifying development and deployment of IP Telephony gateways. Fusion's field-proven hardware components and industry-leading APIs minimize programming requirements and maximize flexibility.

Fusion Hardware

Fusion consists of three hardware components that occupy two ISA slots in a standard PC: an Alliance Generation T1 or E1 (AG-T1 or AG-E1), an AG realtime daughterboard, and a TX Series board.

The AG-T1/E1 provides 24/30 ports of processing for voice and fax plus a full T1/E1 digital PSTN network interface (including PRI ISDN). An 8-port AG-8 can be used in lower-capacity configurations.

The AG Realtime/2 daughterboard, which attaches to the AG-T1/E1, provides real-time vocoding for ports on the baseboard. Using the MVIP bus, the combination of the AG-T1/E1 and AG-Realtime/2 pass traffic to Fusion's third hardware component, a TX2000 or TX3000 board.

The TX Series boards support integration of encoded speech with an Ethernet LAN, which can be an Internet or intranet connection. The TX Series board converts encoded speech to IP packets and supports IP routing and data protocols. The TX boards are available with a wide variety of common data communications interfaces, including 10Base-T and 10Base2 Ethernet.

These three boards, applied within two ISA bus slots, support a full T1/E1 of IP Telephony capability without taxing the host CPU. Because the processing takes place on the boards, multiple Fusion board-sets can be installed in a single chassis, supporting up to four T-spans or E-spans of IP telephony with no degradation of performance and no increase in latency. This allows OEMs, systems integrators, and VARs to create the highest capacity in the industry for a standards-based IP telephony gateway.

Fusion Software

The Fusion software development kit consists of multiple APIs:

- ✔ **Telephony API** (CT Access) for gateway call control and voice processing
- ✔ **Switch Service API** (CT Access) for interconnection of telephony and IP network resources
- ✔ **Vocoder API** (CT Access) for control of speech encoding
- ✔ **Packet Network API** (TX Series Libraries) for control of IP network protocols
- ✔ **H.323 API** for IP Telephony call control.

When Fusion applications receive incoming calls, they spawn caller threads and use CT Access to perform the following tasks: application initialization, port initialization, calls control, event processing and error handling, and parameter management. Additionally, CT Access Switch Service provides a way of making, breaking, and controlling the MVIP connections between Fusion boards. Developers can use the switch service to permanently, or dynamically create data pathways between the telephone line interface and the AG Real-time daughterboard and TX Series Ethernet interface.

Packet Network Integration

The TX control interface is composed of six APIs and a Communication Processor Interface (CPI) library. The CPI library provides a conduit for communicating with the TX series board. The APIs provide functions that simplify control of TX features, such as Virtual Port communication, RTP/RTCP, and embedded UDP routing.

Broadest Selection of Vocoders

Fusion supports the broadest range of vocoders for maximum flexibility of gateway deployment. In addition to the ITU G.723.1 and G.729A algorithms, Fusion also supports Microsoft-GSM and the widely deployed VoxWare MetaVoice RT24 algorithm, which is utilized by Microsoft in NetMeeting client software and by NetScape in the Conference client software. Fusion also provides an open vocoder platform for easy porting of other algorithms as they gain market share or are approved as standards. A CT Access Realtime TRAU library provides a way to start and stop vocoders and a way to manage AG Realtime processes at the board level. The vocoders are combined with integrated tail-end echo cancellers on the AG Realtime board.

H.323 Protocol

H.323 is a broad standard from the ITU that sets specifications for audio, video and data communications over IP-based networks that do not provide guaranteed QoS. H.323 also specifies a series of vocoders to guarantee interoperability among gateways and clients from different vendors. The IMTC has chosen H.323 as the basis for the Interoperability Agreement V1.0 for IP Telephony gateways.

Fusion's H.323 support includes H.225 and H.245. H.225 specifies the syntax and semantics for negotiation at the start of and/or during communication. H.245 specifies media packetization and call setup. For applications that have unique protocol requirements, other stacks may be substituted into Fusion's software architecture. Fusion's software architecture embeds RTP/RTCP on the TX Series board and keeps the remaining H.323 functionality on the host. This partition enables low end-to-end latency and relieves the host from processing real-time audio packets.

RTP/RTCP Library

RTP and RTCP are the accepted H.323 standards for passing real-time data streams over an IP network. Fusion's RTP API provides applications with low-level control over connections that pass real-time data between a circuit-switched network and an IP-based packet network. All RTP connections are initiated, monitored, and eventually terminated via the RTP API.

The RTCP API allows an application to receive RTCP-related information. An RTCP monitor task may be developed for collecting RTCP statistical information on the host. This will allow for QoS monitoring during a session and provide a mechanism for collecting session-specific billing information.

User Datagram Protocol

Within the IP stack, the UDP provides "unreliable services". Fusion supports UDP on the TX Series board. In addition, a host can view the TX Series board as a simplistic NDIS Ethernet board, enabling simultaneous use of the TX board as an intelligent UDP/RTP device and as a host-controlled network interface board. This means that host-based TCP-UDP protocol stacks can be used for command and control events for gateway call-session establishment.

Programmable Jitter Buffer

A unique feature of Fusion is its programmable jitters buffer. As voice packets are transferred across a network without guaranteed QoS, they may be lost or arrive out of sequence. A jitters buffer collects incoming packets and enables Fusion to rearrange the packets into the correct order or to smooth over lost packets. The size of the jitters buffer is configurable on a per channel/session basis, offering a unique feature to control latency for real-time, interactive voice conversations.

Vocoders Supported

The vocoders supported by Fusion include: G.723.1, G.729A, Microsoft GSM 6.10, and VoxWare MetaVoice RT24.

Protocols Supported

The protocols supported by Fusion are: TX 2000/3000, UDP (on TX board), and RTP/RTCP.

Motorola's VIPR

Motorola VIPR is a powerful line of IP telephony products that enable real-time voice and fax communications over the Internet or private intranets. VIPR is a new part of the standards-based MPRouter and Vanguard, Motorola's market-leading family of network accesses devices.

VIPR enables voice and fax traffic to move off the telephony infrastructure and onto the IP data network. By combining data, voice and fax, expensive long distance toll charges and the need for dedicated voice circuits are eliminated, reducing or even eliminating unnecessary charges for data, voice and fax.

The possibilities for implementing value-added service for a network's users are great. Motorola's VIPR technology enables an easy move to VoIP, being totally transparent to network users. Applications of Motorola's VIPR include:

- ✔ Cost savings on long distance calls, as they can be placed over the Internet or Intranet, eliminating long distance charges
- ✔ Network access for traveling and telecommuting employees who can dial into the network to check voice mail, dial another location, or call an extension for the cost of a local connection
- ✔ Optimal bandwidth use by minimizing network traffic and maximizing performance.

Together the MPRouter and Vanguard families provide the industry's most advanced voice and fax support. These versatile network access devices also offer comprehensive support of Frame Relay, ISDN, LAN routing, legacy data protocols, WAN bandwidth optimization and worldwide support for either digital or analog PBX/PABX features. Since all VIPR products are remotely upgradeable and based on a high-performance DSP platform, they are ready for future technologies such as RSVP and RTSP, when they become available. In addition, Motorola's VIPR line is also H.323-compliant and interoperates with Microsoft NetMeeting. These advanced features and the commitment to standards-based implementation makes Motorola VIPR the smartest way of bringing voice and fax to IP data networks.

A summary list of features and benefits of Motorola's VIPR includes:

- ✔ Real-time voice and fax over IP networks
- ✔ Motorola Optimized 8K and 16K CVSELP voice compression provides high-quality voice
- ✔ Digital Speech Interpolation (DSI) is used to minimize wasted bandwidth
- ✔ Voice Switching using DTMF digits
- ✔ Built-in echo cancellation eliminates the need for expensive external hardware
- ✔ Cost-effective concentration for T1/E1 PBX trunks
- ✔ Fax demodulation at 9,600 or 4,800 bps
- ✔ Integrated fax/voice on same port
- ✔ Dedicated DSP hardware ensures maximum compression
- ✔ Lowest price per port
- ✔ FXS, E&M, server card
- ✔ Easy to install and upgrade into existing units.

Nuera Communications' Access Plus Series

Nuera Communications is a leading provider of high-quality packet voice communications equipment and technology for voice/data/image networking over TDM, frame relay and IP networks. Nuera is also a member of the Frame Relay Forum and a co-founder of the VoIP Forum, with great expertise in voice coder development and standards implementation, packet-network optimization, call processing, and network design, installation and support. Nuera also delivers industry-leading DSP-based solutions that optimize bandwidth utilization while maintaining the highest standards of quality for voice traffic and maximum transparency for other applications including fax, signaling, voice-band data and video. Figure 9.1 is a screen shot of Nuera's Web site at URL **http://www.nuera.com**.

Nuera serves carrier, corporate and OEM customers requiring superior voice quality and advanced call-processing capabilities. Nuera is known for its high-quality voice compression at any given rate, as well as its systems for voice/data/fax integration between remote sites. Over 25,000 Nuera systems with over 200,000 voice fax channels are in operation worldwide. In addition, Nuera offers voice compression algorithms including the latest industry standards as well as enhanced, proprietary technology delivering even better quality at very low bit rates. Nuera's voice technology provides high-quality communication at rates from 4.8 to 32 kbps.

TIP

For more information, contact Nuera Communications, Inc. 10445 Pacific Center Court, San Diego, CA 92121; Tel: 619-625-2400, Fax: 619-625-2422.

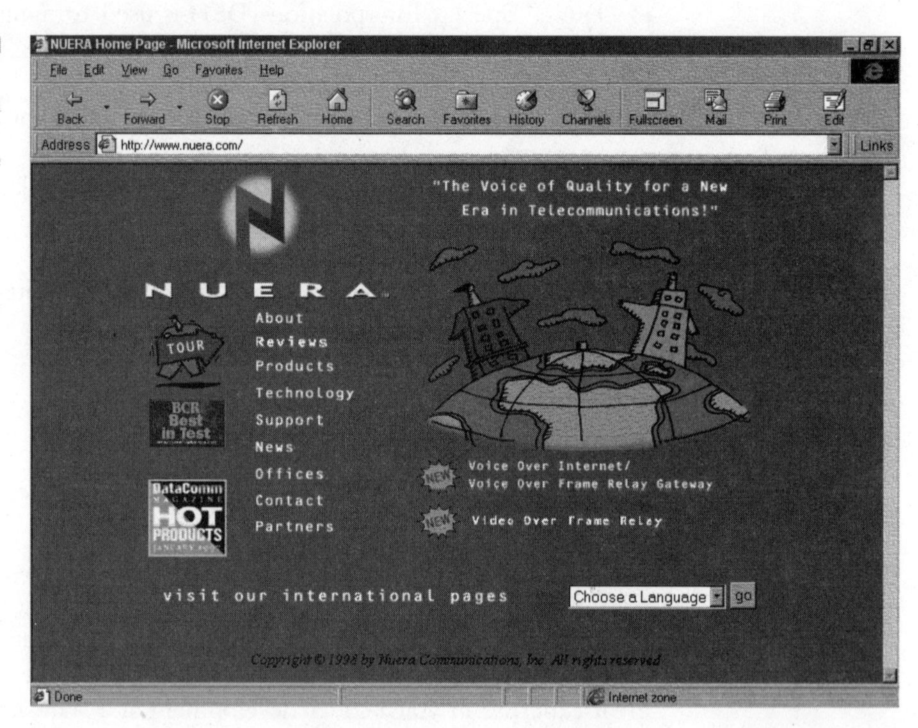

FIGURE 9.1 Nuera Communications' Website

The Access Plus F200

According to Nuera's specifications, the Access Plus F200 is the highest-performance voice over FRAD featuring the industry's best voice quality with Enhanced-CELP (E-CELP) and ITU-standard voice compression. The F200 offers integral switching capabilities and advanced traffic management features, and delivers full T1/E1 throughput on multiple high-speed trunks or data ports. The unit includes analog and digital (T1/E1) interfaces to PBXs or PSTNs and offers capacity for up to 30 voice/fax channels per unit.

The Access Plus F120

The Access Plus F120 provides a voice over frame relay system, outperforming all eight other systems in audio quality, delay and bandwidth utilization in independent National Software Test Labs testing completed in September 1996. The F120 provides analog and digital (T1/E1) interfaces to PBXs or PSTNs and offers capacity for up to 30 voice/fax channels per unit. The product uses award-winning Nuera E-CELP voice compression and packet optimization, plus ITU-standard compression algorithms.

Advanced Voice Compression

The Access Plus F120 Frame Relay Access Device provides the highest-quality voice compression technology in the industry, with complete selection of ITU standard algorithms including G.726, G.728 and G.729 E-CELP algorithms operating at 4.8, 9.47 and 9.6 kbps provide industry-leading quality at each rate. Its adaptive silence suppression substantially minimizes bandwidth usage during speech breaks and maintains natural-sounding conversation, while its integral echo canceller adapts from 0–49 milliseconds to ensure consistent voice quality for all calls.

Access Plus has a unique voice-frame packing feature that optimizes bandwidth over cell-based backbones or low-speed access lines; low-delay voice encoding and adaptive jitter buffer minimize end-to-end speech delay while asymmetric fax channels minimize return-path bandwidth usage. The product is compliant with FRF.11, the Voice over Frame Relay Implementation Agreement.

Call Routing

The F120 provides complete call processing and switches each call independently. Some of its features are:

- ✔ It minimizes implementation and recurring operations costs
- ✔ It reduces the number of ports needed
- ✔ It eliminates tandem calls on the network
- ✔ Calls can be routed through F120s so the voice network can be fully meshed at minimum cost
- ✔ Call routing can control the bandwidth used by voice calls to guarantee data performance
- ✔ It uses clear channel CCS mode for ISDN.

Call Processing

The F120 builds on the voice-switching capability by providing complete call processing and digit translation, as shown in Figure 9.2:

- ✔ Complete end-to-end flexibility independent of voice interfaces
- ✔ E&M analog interface ports can connect to digital FXO interface ports directly without conversion equipment
- ✔ Digit translation provides a unified dialing scheme capable of 20-digit translation and 40-digit outpulsing.

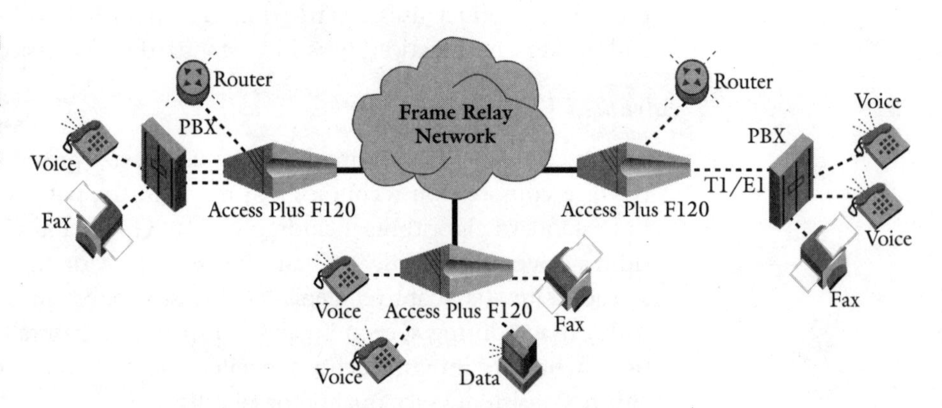

FIGURE 9.2 Nuera's F120 is built on voice switching technologies

Data FRAD and Switch

The F120 is both a data/voice FRAD and a frame relay switch, very easy to integrate into existing data networks, and supporting both the DTE and DCE sides of the UNI interface. It functions as a frame relay network to attached devices including routers and front-end processors.

Digital Interfaces

The F120 makes installation and expansion easy by providing direct T1/E1 interfaces for voice ports. It also eliminates unnecessary digital-to-analog conversions and subsequent distortion, and simplifies tuning of the voice ports so that the volume levels are more consistent across all ports.

Access Plus F200ip

The Access Plus F200ip is an integrated voice over Internet/voice over frame phone-to-phone gateway, bringing voice quality to Internet applications by delivering IP voice via Ethernet and frame voice via high-speed serial trunks.

A High-Performance System

The F200ip provides flexibility in network protocols for optimizing quality/cost tradeoffs:

- ✔ It runs both voice over IP and voice over frame relay concurrently
- ✔ It can switch voice calls from any port to any port over any protocol
- ✔ It has simple LAN installation and provisioning when using IP.

An Advanced Voice Compression

The Access Plus F200ip provides the highest quality and widest range of voice compression technology in the industry:

- ✔ ITU G.728 LD-CELP, G.729 CSA-CELP, and G.726 ADPCM standard algorithms from 8 kbps to 32 kbps
- ✔ E-CELP advanced proprietary algorithms operating at 4.8, 9.47 and 9.6 kbps provide industry-leading quality at each rate
- ✔ Low-delay voice encoding and adaptive jitter buffer minimize end-to-end speech delay
- ✔ Integral echo canceller adapts from 0–49 millisecond to ensure consistent voice quality over long tail circuits
- ✔ Sophisticated lost-packet recovery methods help maintain consistent quality in harsh environments
- ✔ Voice compression rates are negotiated to provide configuration flexibility
- ✔ There is modem transparency over voice channels.

High Bandwidth Efficiency

Another unique feature of the F200ip is its incorporated technology, designed to minimize WAN bandwidth. Features include:

- ✔ Programmable voice packeting improving bandwidth efficiency
- ✔ Adaptive silence suppression minimizing bandwidth usage during speech breaks and maintaining natural-sounding conversation
- ✔ Asymmetric fax channels minimizing return path bandwidth usage
- ✔ Data fragmentation is configurable on a DLCI basis to minimize delay of voice traffic over low-speed interfaces.

Call Routing

The F200ip provides complete call processing by switching each call independently with full-digit and interface translation. Below are some of its characteristics:

- ✔ It minimizes the number of ports needed
- ✔ It eliminates tandem calls on the network
- ✔ Call switching allows the voice network to be fully meshed at minimum cost
- ✔ Digit translation provides a unified dialing scheme capable of 20-digit translation and 40-digit outpulsing
- ✔ There is complete any-to-any connectivity, independent of voice interfaces
- ✔ It generates call-detail records for per-call billing.

Flexible Voice Interfaces and SNMP Network Management

The F200ip provides direct T1/E1 interfaces for voice ports, up to 24/30 channels. The T1/E1s eliminate digital-to-analog conversions and subsequent distortion. F200ip also has analog interfaces programmable with a wide range of options—FXO, FXS, E&M.

A Brief Overview of NueraView

NueraView has a robust configuration, statistics, diagnostics, and alarm management. Its database assures that configuration information is safely backed up. The NueraView SNMP network management system (NMS) is a powerful system which allows the network operations and staff to manage networks with minimum effort. The graphical user interface provides advanced NMS functions such as configuration, monitoring, diagnostics, statistics collection, and alarm collection. NueraView simplifies network management while boosting performance across multi-node networks with voice/fax/data/video traffic, as shown in Figure 9.3.

Increased Network Uptime

By providing instantaneous alarms graphically displayed on a network map, NueraView quickly helps isolate faults demonstrating:

- ✔ **Faster fault resolution**—Graphical alerts highlight problems as they occur
- ✔ **Easier restoral**—Configuration database is used for system replacements
- ✔ **Fewer configuration errors**—Pull-down menus make typing unnecessary
- ✔ **Remote dial access**—Full control of network from remote NMS is possible
- ✔ **Consolidated/central management**—All devices are monitored on single map.

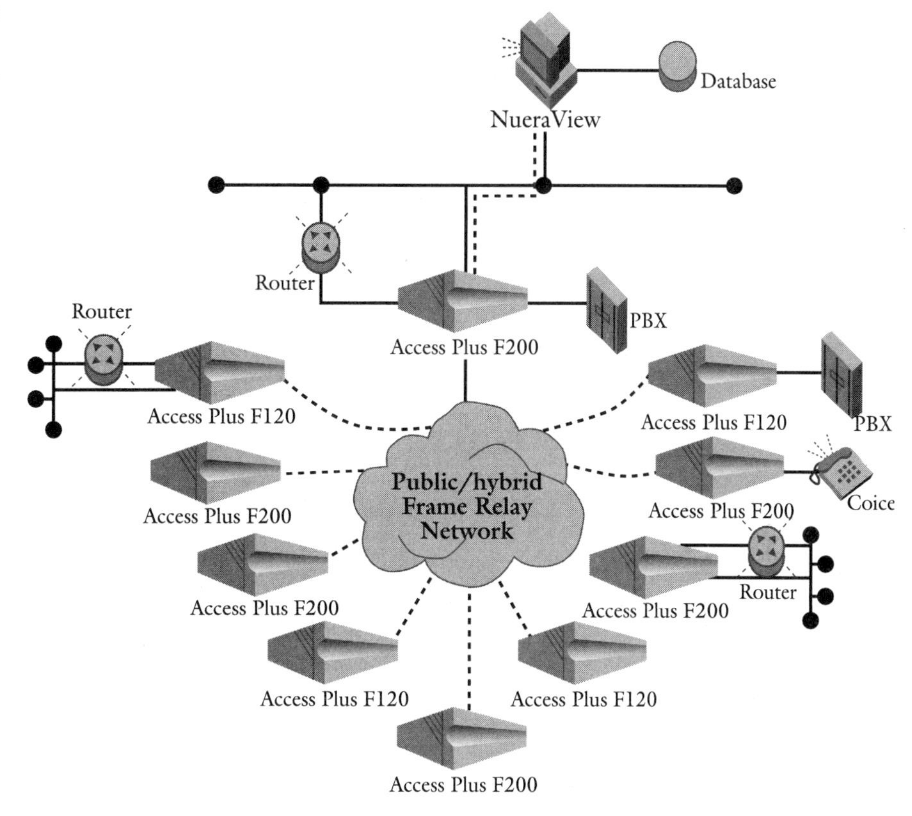

FIGURE 9.3 Network management is simplified through Nuera's NueraView

Easier Operations

NueraView enables configuration time for systems to be minimized (see Figures 9.4, 9.5 and 9.6 for screen shots of its GUI interface):

- ✔ **Point and click to change configuration and to view status and trap logs**—This is faster and more accurate than typing
- ✔ **Hierarchical graphic network map**—Arrange network items logically while maintaining full monitoring capability
- ✔ **Offline node configuration and staging**—Preconfigure nodes using templates
- ✔ **Configuration templates**—All parameter options are readily viewable
- ✔ **Online help and documentation**—Point and click through user a guide
- ✔ **Remote dial access into NMS console**—Monitor a network from anywhere
- ✔ **Code and configuration downloads**—Upgrade code or configurations from a central site.

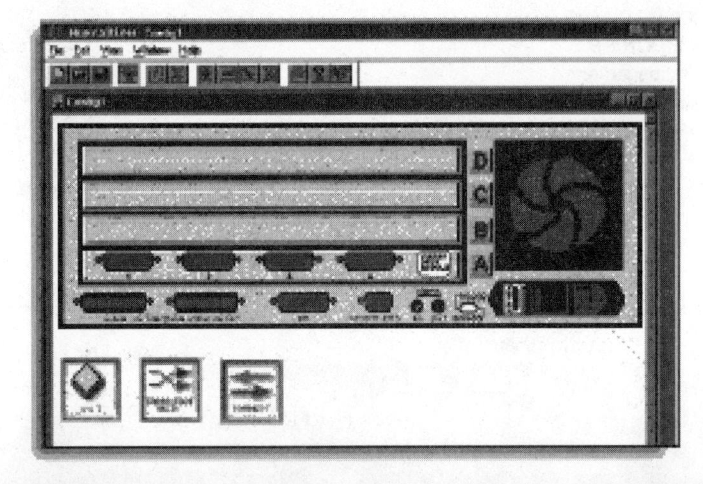

FIGURE 9.4 NueraView's easy-to-use configuration utility

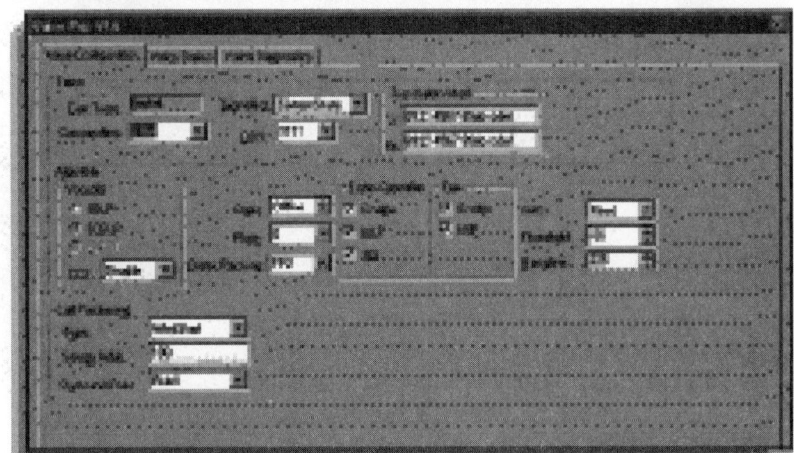

FIGURE 9.5 NueraView leverages the extensive software control provided by the Access Plus F-Series

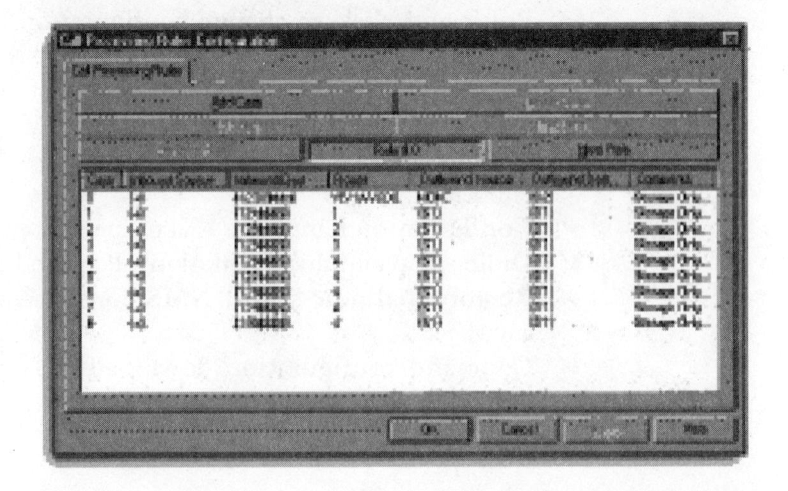

FIGURE 9.6 Dial plans are simple to design and maintain

Portable Voice Module

Portable Voice Module is a OEM package, delivering top-rated voice support to customers' FRADs, routers or other data devices connecting to TDM, frame relay, IP or ATM networks. It also features Nuera's E-CELP and ITU-standard voice compression algorithms plus Nuera's packet-network quality enhancements and call processing features package. It includes DSP modules, telco interface modules and a host interface for the target platform, plus software APIs to facilitate development and expedite time to market.

Access Plus 100 and Access Plus 200

Access Plus 100 and Access Plus 200 are both TDM-based voice compression and voice/fax/data integration and bandwidth management systems. They feature Nuera's E-CELP compression plus ITU standard algorithms, analog and digital voice interfaces and capacity for up to 30 voice/fax channels per system as well as point-to-point and point-to-multipoint capabilities.

Clarity Series CS8000—Voice/Data Multiplexing

Clarity Series CS8000 provides TDM-based voice compression and voice/fax/data integration for up to eight analog voice/fax channels, and is engineered to enhance network performance by (see Figure 9.10 on page 251):

- ✔ Integrating multiple channels of voice, fax and data onto a low-cost digital circuit
- ✔ Offering advanced voice card hardware featuring completely programmable interfaces
- ✔ Easy configuration and proven system reliability producing "install-and-forget" wide-area networking
- ✔ Low-delay data transmission rates (9.6 kbps to T1) supporting a wide range of devices including bridges, routers, and video conferencing equipment.

Voice Compression

The CS8000 provides delivery of consistent high-quality communication across a broad selection of compression algorithms:

- ✔ CELP algorithms operating at 5.3, 8, and 9.6 kbps provide outstanding voice quality at each rate
- ✔ Adaptive Transform Coding (ATC) compression rates are completely variable between 7.47 kbps and 32 kbps

- ✔ Integral echo cancellation (G.165 compatible) in excess of 32 msecs ensures voice quality for off-net calls
- ✔ Fax relay support provides transparent fax operation at 2.4, 4.8, 7.2, and 9.6 kbps.

A wide range of connectivity solutions allows each location to be tailored for the lowest-cost configuration. Flexible allocation of bandwidth provides for numerous voice, data, and video services on a T1 local loop interface. Advanced functionality over satellite, sub-rate or T1 circuits allows users to build optimum network designs.

Qwest Communications' Macro Capacity Fiber Network

Qwest Communications International, Inc. is a digital-age communications company building a high-capacity, fiber optic network. With its cutting-edge technology, the Qwest Network is able to provide high-quality data, video and voice connectivity with uncompromised security and reliability to businesses, consumers and other communications service providers.

Qwest constructs and installs fiber optic communications systems for interchange carriers and other communications entities, as well as for its own use. Qwest is expanding its existing long distance network into an approximately 16,000 route-mile coast-to-coast technologically advanced, fiber optic telecommunications network. The network will connect approximately 125 metropolitan areas that represent about 80% of all long distance traffic in the United States.

Through a combination of the Qwest Network and leased facilities, the company will continue to offer interstate services in all 48 contiguous states. The Network will connect to international cable heads for transatlantic and transpacific transmission to Canada and Mexico. Qwest recently extended its network to the United Kingdom through an exchange of capacity for two 155-megabit circuits that will carry international data and voice traffic between London and New York. The company also is extending its network approximately 1,400 route-miles into Mexico through dark fiber to be owned by Qwest on the fiber optic system of a third party.

As the demand from interchange carriers and other communications entities for advanced, high-bandwidth voice, data and video transmission capacity increases, due to regulatory and technical changes and other industry developments, Qwest strategically positions itself to provide the products and service this high bandwidth demands. The company is also committed to addressing the changes it generates. These anticipated changes and developments include:

- ✔ Continued growth in capacity requirements for high-speed data transmission
- ✔ ATM and Frame Relay services
- ✔ Internet and multi-media services and other new technologies and applications
- ✔ Continued growth in demand for existing long distance services
- ✔ Entry into the market of new communications providers
- ✔ Reform in the regulation of domestic access charges and international settlement rates, which Qwest expects, will lower long distance rates.

Qwest's Macro Capacity Fiber Network

The Qwest Macro Capacity Fiber Network is designed to be the highest-capacity digital infrastructure in the world. With wave-division multiplexing and advanced optical transmission technology, it transmits with unprecedented speed, reliability and affordability. By making available the following advantages, Qwest's state-of-the-art infrastructure can very likely become the communications backbone of digital America:

- ✔ **Quantum capacity**—Designed with a transmit capacity of up to two terabits per second, the Qwest Macro Capacity Fiber Network can carry more information than any other United States communications network
- ✔ **Hyper speed**—Video, images and data (including VoIP applications) can be sent from coast to coast in the blink of an eye. At full capacity, the Qwest Macro Capacity Fiber Network will transmit two trillion bits of multimedia content per second—or the entire Library of Congress across the country in 20 seconds.
- ✔ **Absolute data integrity**—The Qwest Macro Capacity Fiber Network is designed to offer unprecedented protection against data loss. Its Absolute Data Integrity is the new world standard for error-free transmission. The network's non-zero, dispersion-shifted glass, pure OC-192 transmission technology and advanced SONET ring architecture ensure less than one bit of error in every quadrillion bits. That is the equivalent of one grain of sand out of place on a 20-mile stretch of beach. Consequently, this advantage of the Qwest Network sets a new world-standard precedent of reliability and accuracy in data transmission.
- ✔ **Superior service reliability**—The sophisticated fiber optics of the Qwest Macro Capacity Fiber Network are encased in a thick plastic protective conduit buried in a highly secure environment, principally along railroad lines. Moreover, the fiber is laid out in an advanced SONET ring architecture that provides automatic, instant re-routing should disruption of any kind occur.

- ✔ **Low cost position**—The Qwest Macro Capacity Fiber Network's advanced fiber and transmission electronics provide the company with lower installation, operating and maintenance costs than do older fiber systems typical in commercial use today. In addition, Qwest has entered into construction contracts for the sale of dark fiber along the route of the Qwest Network which will reduce Qwest's net cost per fiber mile with respect to the fiber it retains for its own use. As a result of the cost advantages, Qwest's low cost position will enable the capture of market share and take advantage of the rapidly growing demand for data transmission, multimedia and long-haul voice capacity.

The Qwest Macro Capacity Fiber Network is paving the way for 21st century communications. Through secure, reliable and affordable data, image and multimedia content transmission, Qwest will accelerate the expansion of the digital age in America.

Already, Internet service providers and other telecommunications companies are using the capacity, speed, affordability and flexibility of the Qwest Network to upgrade and expand their offerings. Businesses of all sizes are turning to Qwest to help them unlock their future potential through Qwest's provision of long distance, IP and data transmission services. As Qwest expands across the nation, people and businesses everywhere will discover digital shopping, digital entertainment, digital education and digital communication as a result of Qwest's superior Macro Capacity Fiber Network.

The Qwest Network enables advanced digital communications to seem as simple as a telephone call. Services are easy to use, easy to change and easy to upgrade. Qwest even provides capacity on demand—customers can expand or shrink capacity on their own in an instant—as well as a broad portfolio of voice and data services that give all customers the choices they need to bring them into the 21st century.

By early 1999, the Qwest Network will be operating in more than 125 United States cities, with extensions into Mexico, connections to the United Kingdom and other international networks yet to be determined.

For more information on Qwest and their products, contact corporate headquarters: Qwest Communications International Inc., Qwest Tower, 555 Seventeenth Street Denver, CO 80202. Tel: 303-291-1400 or 800-899-7780, Fax: 303-291-1724, or on the Web at **http://www.qwest.net**.

NetSpeak Corporation's WebPhone

NetSpeak Corporation is one of the leading developers and marketers of IP telephony technology providing business solutions for concurrent, real-time interactive voice, video and data communications over packetized data networks such as the Internet, LANs and WANs. NetSpeak Solutions allow organizations to build new voice and video-enabled communications networks or to add these communications capabilities to their existing enterprise.

With VoIP picking up, IP telephony is receiving increasing attention from the telecommunications, networking, software and investment communities. It is estimated that US companies spent $83 billion on long distance calls last year. According to researchers, IP telephony has the potential to enable companies drastically to reduce their telecommunications costs. But more important, IP telephony offers the ability for multimedia communications—voice, fax, data and video over a single channel. Although cost savings are currently generating the demand for IP telephony, ultimately the ability to provide multimedia communications and enhanced user services will likely be the key growth driver of the IP telephony industry.

NetSpeak provides a complete suite of reliable, real-time, high-performance multimedia communications systems over packetized data networks for service providers, businesses, call centers and consumers. All NetSpeak solutions utilize its patent-pending virtual circuit-switching technology to dramatically enhance the multimedia capabilities of both private and public IP-based networks, and drastically reduce the cost of providing advanced services over these networks.

WebPhone

NetSpeak's WebPhone 4.0 is the latest version of its Internet telephony software. The product gives users the ability to have voice, video and data communications over the Internet and other TCP/IP-based networks through an IP telephony software package. WebPhone extends communication into the realm of multimedia by combining audio, video and text capabilities—without the cost of long distance phone calls.

NetSpeak has licensed elemedia, Lucent Technologies' G.723.1 speech coder implementation, for use in all NetSpeak client applications including WebPhone 4.0. NetSpeak has also licensed DSP Group, Inc.'s patent rights for G.723.1. elemedia and DSP Group both represent separate sets of intellectual property ownerships.

The G.723.1 speech technology standard is part of the H.323 specification adopted by the ITU as the international standard for voice over packet-switched networks such as the Internet and LANs. G.723.1 is also selected by the VoIP Activity Group, part of the International Multimedia Teleconferencing Consortium (IMTC), as the preferred speech coder for Internet telephony over modem connections.

By complying with H.323, multimedia products and applications from multiple vendors can interoperate, allowing users to communicate without concern for compatibility. The incorporation of the G.723.1 technology will give WebPhone users the highest quality industry-compliant capabilities while providing critical standard components for interoperability.

WebPhone 4.0 is a full-featured IP telephony software package, extending PC-to-PC communications into the realm of multimedia by combining voice, video and text capabilities. In addition to interoperability support, WebPhone 4.0 offers Audio Setup Wizard for configuring speakers and microphone; it is easy and fast to install. WebPhone also provides inbound and outbound activity logging of calls, an enhanced user guide and recall for the last five parties called.

The features provided by WebPhone include:

- ✔ Point-to-point voice and video over the Internet or any TCP/IP-based network
- ✔ Real-time, full-duplex voice communications
- ✔ State-of-the-art cellular phone interface for ease of use
- ✔ H.323 support; communicates with any H.323-compliant Internet telephone
- ✔ Audio Setup Wizard for configuring speakers and microphone
- ✔ Voice Auto Detection which automatically detects and adjusts to the voice, improving audio quality
- ✔ Online/offline integrated voice mail system
- ✔ Four lines with call holding, muting, do-not-disturb and blocking options
- ✔ Complete Caller ID information
- ✔ Speed dial, redial, call conferencing and call transferring
- ✔ Recall for last five parties
- ✔ Video phone support using the H.263 standard
- ✔ Fast video frame delivery for low- and high-bandwidth connections
- ✔ Large video display area with self and remote views
- ✔ Interactive party-specific TextChat
- ✔ Inbound and outbound activity logging
- ✔ Integrated real-time information assistance
- ✔ Personal directory to store frequently called parties

- ✔ Audio quality enhanced with TrueSpeechTM, G.723.1 and GSM voice compression
- ✔ Account information and dialing parameters for use with NetSpeak gateway products.

The system's requirements include:

Hardware
- ✔ 90 MHz Intel Pentium
- ✔ 16 MB RAM
- ✔ 10 MB available (free) hard disk space
- ✔ MCI-compliant sound card that supports 8 kHz or 11 kHz sampling
- ✔ VGA card capable of displaying 256 colors

Hardware for Video
- ✔ 120 MHz Intel Pentium
- ✔ 28.8 Kbps connection
- ✔ Camera or other input source
- ✔ Capture card if required by camera or other input source
- ✔ Operating System
- ✔ Microsoft Windows 95 or Windows NT
- ✔ Network connection
- ✔ 28.8 Kbps connection
- ✔ Windows Sockets version 1.1 or later
- ✔ TCP/IP network connection (LAN, WAN, or Internet).

TIP

Download WebPhone free before you buy it at the URL below.

For more information on NetSpeak, write 902 Clint Moore Road, Suite 104, Boca Raton, FL 33487-2846, or call (561) 998-8700, or fax (561) 997-2401. Their Web site is at **http://www.netspeak.com/**.

NetPhone's PBX Servers

NetPhone, Inc. is a privately held company headquartered in Marlborough, MA. NetPhone designs, manufactures and markets intelligent telephone systems and telephony applications for medium- and-small size offices. NetPhone PBX Servers are a new generation of telephone systems that integrate the power, ease-of-use, and economics of PC servers with the familiarity and reliability of the telephone. These open, reliable telephone systems allow organizations easily and affordably to deploy a wide variety of computer telephony applications to increase sales, enhance customer service, reduce telephone costs, and boost staff productivity.

Large corporations have adopted Computer Telephony (CT) as a strategic communications technology. CT has unfortunately not been an option for smaller companies, because it was either unavailable from a telephone system provider or prohibitively expensive.

NetPhone PBX Servers, enables CT solutions for smaller business by making computer telephony-enabled PBX systems cost-effective by incorporating all components of a complete PBX, and server software that implements all standard computer telephony APIs. The PBX boards simply plug into standard PC servers running Windows NT or NetWare. Telephony applications supplied by NetPhone and third parties execute on the PBX Servers and PC clients. NetPhone PBX Servers and applications deliver the productivity enhancements of PC-based visual phone control with caller ID, database screen pops, phone management, visual voice mail, and more.

NetPhone PBX boards are compatible with industry-standard application interfaces including Microsoft's TAPI and Novell's TSAPI, and provide a series of benefits, itemized below:

- ✔ **Advanced PBX Call Processing**—Provides call waiting, call hold, call transfer, call conferencing, call forwarding, call pick-up, call group covering, call group hunting, and call queuing telephony services for office and call center environments
- ✔ **PhoneMaster**—This Windows-based desktop call control application features caller-ID handling, screen pops of desktop applications, control and status of inbound and outbound calls, and a graphical interface to voice mail
- ✔ **VoiceMaster**—This comprehensive voice mail manager allows users to record, access, manipulate, and forward messages from their telephones with standard DTMF touch-tone commands
- ✔ **Flexible Auto Attendant**—This supports customized greeting messages for incoming callers
- ✔ **Customizable Application Templates**—These are templates for horizontal and vertical market telephony applications such as ACD, advanced call filtering, account/order status and inquiry, and many additional IVR applications
- ✔ **High Availability Design**—This patent-pending PBX switch architecture eliminates dependency on host-resident switch call control software, ensuring the telephone system continues to operate even when the host PC server is taken off-line
- ✔ **Scalable Capacity of User Extensions and Trunks**—These enable an organization's phone system to expand easily as their business grows

- ✔ **Support for Standard Phone Sets and Speaker Phones**—This eliminates the need for businesses to spend money on proprietary, highly expensive telephones
- ✔ **Simplified Telephone System Administration**—This features a Microsoft Windows GUI that provides local and remote administrators a point-and-click interface to handle telephone moves, adds and changes.

NetPhone PBX servers are based entirely on industry standards. The company is a member of the Voice-Over-IP Forum. As a result the servers are supported by a wide range of third-party computer-telephony applications, including those from Microsoft, CallWare, CoreSoft, Brooktrout, Tobit Software, Decisif, SoftLinx and many others.

TIP

For additional information about NetPhone or their products, write them at 313 Boston Post Road, West Marlborough, MA 01752. Tel: 508-787-1000, or Fax: 508-787-1030. E-mail: **info@netphone.com**.

Vocaltec's Internet Phone and Telephony Gateway Server

VocalTec Communications Ltd. is another Internet telephony company. An Israel-based firm, VocalTec develops and markets award-winning software that enables voice, fax and multi-media communications over packetized IP networks, the Internet and corporate intranets. The company also develops open systems to bridge the Internet/intranets to the PSTN.

VocalTec pioneered the Internet telephony market with the introduction of Internet Phone in 1995. For the first time, anyone who owned a multimedia PC could make or receive a call from a computer anywhere in the world with no long distance charges.

VocalTec continues to drive Internet telephony's evolution into the mainstream by being first to market with customized client/server solutions for the corporate and carrier markets. These products offer corporations increased productivity and cost savings and enable traditional and new generation telcos, such as Internet telephony service providers (ITSPs), to provide more affordable and powerful communications services to customers.

Vocaltec plays a central role in the development of interoperability standards. During 1996, the company demonstrated its commitment to the standards process by working closely with Microsoft and Intel to achieve interoperability between VocalTec Communications' Internet Phone software and Microsoft NetMeeting and Intel's Internet Phone software by announcing their support of the International Telecommunications Union H.323 open standard. The Company has also announced its founding membership in the VoIP Forum. VocalTec was also the first company to ship a product using the UDP/IP protocol (one of the building blocks of Internet connectivity) for real-time voice communication.

VocalTec and Cisco co-founded the VoIP Forum of the IMTC in May 1996 to ensure and promote industry-wide interoperability of Internet voice communications products. Within the VoIP forum, VocalTec Communications has been developing its Call Management Agent (CMA) technology, expected to be extremely important to the development of truly rich and useful IP-based telephone products.

VocalTec Communications has also been active in the development of the emerging ITU H.323 system. Recognizing that the real promise of IP-based telephony lies in the added functionality of new features, including integration with Web-based voicemail, multi-party collaboration, and multimedia supplementary services (call waiting, call transfer, etc.), the company places a high level of importance on enabling all of these new features in a standard way so that the customer can be certain that the system works as a whole.

The VocalTec Ensemble Architecture

The VocalTec Ensemble Architecture (VocalTec EA) is an open standards-based software platform, which forms the foundation for IP communications solutions from VocalTec. The design and architecture of VocalTec EA relies on years of VocalTec field experience in IP communications and the company's active role in Internet telephony standards organizations. VocalTec EA is a third-generation architecture capable of sustaining widescale deployment of IP communications in the corporate and service provider environments.

The Internet Phone

The Internet Phone is Vocaltec's client software that enables users simultaneously to talk to and see each other in real time for the cost of an Internet connection. Internet Phone Release 5 has many new, improved features including enhanced audio and video, support for international standards, PC-to-standard phone calling and a new community browser. In addition to being a standalone product, Internet Phone is a key component of VocalTec's corporate and carrier Internet telephony client/server solutions.

The VocalTec Telephony Gateway Server

The VocalTec Telephony Gateway Server bridges the gap between the traditional telephone network and the Internet/intranets to enable unlimited long distance calling and faxing. It allows users to connect over the Internet or intranet telephone-to-telephone, PC-to-telephone, telephone-to-PC, fax-to-fax, and Web browser-to-telephone. The Telephony Gateway uses the power of the Internet protocol standard to improve the flexibility and performance of business communications systems while reducing long distance phone charges.

For a full list of Vocaltec's products and specifications, visit their Web site at **http://www.vocaltec.com/about/aboutus.htm** or write or call:

In Israel: 1 Maskit Street, Herzliya 46733, Tel.: +972-9-970-7777, Fax: +972-9-9561-867.

In the United States: 35 Industrial Parkway, Northvale, NJ 07647, Tel.: 201-768-9400 or Fax: 201-768-8893.

Vienna Systems' IP Telephony Solution

Vienna's IP Telephony switching products enable Service Providers to offer new services and enter new businesses. The company designs and manufactures hardware and software products to distribute voice, fax and video calls across IP networks, both corporate (intranet) and public (Internet). The company offers a true end-to-end solution ranging from client products to large-scale carrier gateways.

Vienna's Products

Vienna has a complete suite of IP telephony solutions, enabling service providers to extend new services to residential and business users. A unique combination of server applications, gateways and end-user devices, Vienna's IP telephony product line delivers intelligent voice and fax services over geographically distributed, large-scale voice and data networks (see Figure 9.7).

Vienna's system architecture provides the following key benefits: rapid service deployment, overall network scalability and customized services.

FIGURE 9.7
The Vienna Systems Platform

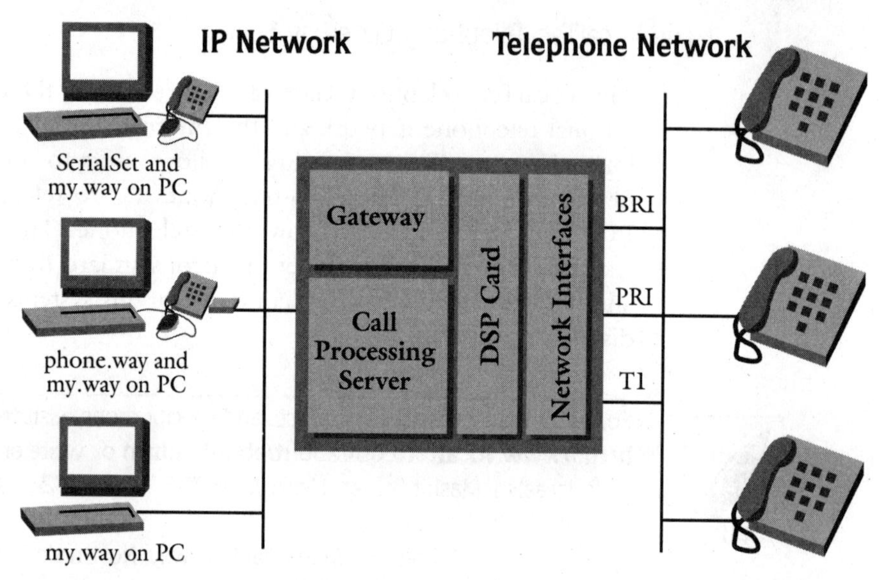

Vienna's products enable service providers to:

- ✔ Offer long distance phone-to-phone service. Vienna gateways are installed at Points of Presence (POPs) in different calling regions throughout a network to provide the connection between the public telephone network and the Internet. Customers simply call a local gateway, enter an authorization number, and dial the destination number.
- ✔ Offer PC-to-phone service. Phone calling capability can be added to an existing suite of services such as high speed Internet access.

Vienna provides the capability to offer real-time fax service with the simple installation of fax modem cards. Fax service can be provided to customers without impact on existing voice quality or network scalability.

Using Vienna's sophisticated solution, service providers can afford corporate customers customized Interactive Voice Response (IVR), dialing plans, network access privileges, authentication, routing and billing. Networks can be "partitioned" among many businesses, creating a virtual private network for corporate customers.

A key component of Vienna's solution is the Vienna Call Processing (CP) Server that extends the traditional PBX voice communication features to voice, data and fax traffic. To provide access to this environment, Vienna gateways provide an interface between the public telephone network and the IP network. The client interface, **my.way**, provides multi-line PBX functionality on the desktop.

Vienna's product family consists of:

- ✔ **CPServer**—The server provides traditional PBX features across the network and delivers multimedia calls across private (Intranet) and public (internet) networks. In addition to providing call-processing features, the server is also responsible for maintaining all configuration and user data.
- ✔ **Vienna Gateway**—The gateway serves as a network interface between an IP network and the public telephone network.

Vienna Desktop Applications

- ✔ **my.way**—This desktop application extends a wide range of multimedia applications to the end-user. It resides on the user's desktop, as shown in Figure 9.8, providing multiline phone and PBX functionality. Up to eight lines are available at the desktop, and users can rely on the corporate-wide telephone directory or create their own for speed dialing. IP address resolution allows users to dial by name or extension number. Users can perform standard functions like answer/hang-up, hold, retrieve, transfer, call forward, DND, mute, pick-up, conferencing and more. Data collaboration is easily launched within the application and editing rights can be shared. my.way is compatible with standard data conferencing, application sharing, and data collaboration applications such as Microsoft's NetMeeting.

FIGURE 9.8 my.way desktop application, by Vienna Systems

- **IPShuttle**—This small, user-installed customer premises device allows standard telephones to exploit the new IP telephony networks and services offered by cable and phone companies. It is an IP telephony device that can be deployed easily on a large scale to residential and business environments. It uses existing subscriber telephones as the actual communications tool and is activated and maintained from a central server.
- **IPCourier**—This fully featured Ethernet telephone allows both business and residential customers to benefit from IP telephony services. While it can work in conjunction with a user's PC for call directories and other telephony functions, IPCourier is not dependent on the PC. It provides a full complement of features and access to the IP network even if the PC is turned off.
- **SerialSet**—This serial telephone connects directly to the PC, as show in Figure 9.9, allowing callers to connect through their computer for voice communication using the my.way application. It features familiar functionality like flash, re-dial and mute. As an alternative to sound cards and microphones, it provides for increased privacy and improved IP voice quality for users on the network. At the office, a single connection for voice and data at the desktop makes moves, adds and changes easier for the network administrator. Its portability means remote workers can connect SerialSet to their laptops to place and receive IP voice calls through dial-up access to the network when away from the office.

FIGURE 9.9 Serial.Set is a serial telephone to be connected to the PC, by Vienna Systems

- ✔ **phone.way**—This serial telephone adapter, shown in Figure 9.10, connects a standard phone to the PC. Vienna's phone.way connects existing standard phones to the PC through the PC's serial port, eliminating the need for a sound card and microphone. Operating with the my.way desktop application, phone.way allows callers the choice of dialing and accessing call-processing features from their PC keyboard or from the telephone keypad. If the PC is powered off or the network is unavailable, phone.way will automatically switch calls to bypass the PC. The desktop telephone will then connect through a second RJ-11 jack to an attached PBX.

FIGURE 9.10 Phone.way serial telephone adapter, by Vienna Systems

The Vienna architecture has been designed to be fully distributed and flexible, making it simple for customers to build new applications and services.

Lucent Technologies' Internet Telephony Server-E (ITS-E)

Lucent Technologies, headquartered in Murray Hill, NJ, designs, builds and delivers a wide range of public and private networks, communications systems and software, data networking systems, business telephone systems and microelectronics components. Bell Laboratories is the research and development arm of the company. For more information about Lucent Technologies, visit the web site at **http://www.lucent.com**.

Lucent is investing heavily in VoIP technology, as the market expands rapidly. At NetWorld+Interop, in May 1998, Lucent announced a portfolio of enhanced VoIP products, including a mixed-media application server and an Internet telephony gateway, that enable enterprises and service providers to create new IP-based applications, effectively scale and migrate their IP-based networks and improve business productivity.

Lucent is also bringing new capabilities to the ADSL market, with its new ADSL access solutions that reside on the customer premises and in the service providers' networks as well as in chip sets for computers.

The Internet Telephony Server – E (ITS-E)

Lucent Technologies offers a Microsoft Windows NT server-based solution that places voice and fax calls over IP networks using voice compression software developed by Bell Labs' new division, elemedia. The solution, known as the ITS-E, works with the DEFINITY ECS and most existing telephone systems and is connected to the PBX via a T1/E1 Tie Line or analog line interface. On the IP network side, the ITS-E is connected via a standard 10/100BaseT Ethernet interface. ITS-E supports calling between two standard telephone sets or two fax machines. ITS-E also supports calls between a telephone and an H.323 standard voice over IP PC program, such as Microsoft's NetMeeting.

The voice over IP using ITS-E is near toll quality. The SX7300 algorithm has been evaluated as having a Mean Opinion Score (MOS) of 3.5 (based on tests performed by COMSAT Laboratories, Clarksburg, MD in August 1995).

MultiMedia Communications eXchange Server (MMCX 2.1)

Lucent's MMCX 2.1 is leading the industry as the first fully standards-compliant H.323 conference server, by extending multiparty media conferencing to a broad array of network connections, including ATM, wireless LANs, remote access and H.320 and H.323 endpoints.

Positioned as an IP-based tool for improving real-time communications in business networks, MMCX 2.1 also adds improved audio and video quality and increased interoperability by enabling UNIX workstations and PC endpoints to be co-resident on the same server.

The MMCX 2.1 support's RADVision's new OnLAN L2W - 323 Multimedia H.323 Gateway, making it possible for MMCX, H.323 and other H.320 clients, such as PictureTel Corp. room systems, to communicate over the same LAN. The gateway also enables MMCX users to tie their existing ISDN-based PC and room videoconferencing systems into the LAN in a cost-effective way.

To help network administrators reduce the amount of multimedia traffic on the WAN, the MMCX 2.1 now supports IP multicasting. In addition, for the first time, enterprises deploying large networks of MMCX servers can use sophisticated network management tools such as Hewlett-Packard's HP OpenView.

The H.323-compliant MMCX client and server software now supports Microsoft Windows PCs, UNIX workstations and telephony-based end-points, enabling customers to establish spontaneous point-to-point or multi-party calls using a variety of voice, video and data.

Companies such as Lockheed Martin Tactical Aircraft Systems of Fort Worth, TX, have begun rolling out the latest version of the MMCX to more than 100 engineers, designers and partners around the world by the end of this year.

Another example of Lucent's solutions using VoIP is achieved through mixed-media conferencing, which will allow virtual meetings to be held, as well as group reviews of avionics software code as developing the joint strike fighter begins.

The joint strike fighter, due in 2008, is a multirole aircraft for the U.S. Air Force, Navy and Marines and the British Navy. Lockheed Martin is teaming up with British Aerospace, Northrop Grumman and others to produce the first demonstration planes. Lucent's MMCX VoIP solution will enable Lockheed Martin to resolve many business issues without having to travel as sharing of data and plan activities can be delivered via MMCX.

Internet Telephony Server for Service Providers (ITS-SP)

The ITS-SP allows service providers to route voice and fax communications over the Internet with near toll-grade reliability and quality. Release 2.0 includes a new network architecture designed to offer key features today, while providing a path for service providers to increase the scalability of their networks, reduce their per-port prices, increase their overall manageability and provide plug-and-play applications.

One of these new architectural enhancements is the ITS-Service Access Manager (ITS-SAM) which allows service providers to create zones for managing multiple gateways. The ITS-SAM permits service providers to set up separate zones—a collection of endpoints, PC clients and gateways—for delivering services such as authentication control, security and call routing. For example, zones for up to 25 gateways and 500 PC clients can be set up per ITS-SAM.

Another component of the ITS-SP 2.0 is the ITS-Administration Manager (ITS-AM); a secure Web-based network management tool that allows service providers to manage multiple gateways from a central location.

The ITS-SP 2.0 also includes features that enable service providers to tailor their level of service with customers through service level agreements. For example, service providers can choose to route calls to an alternative data network or PSTN when the network is congested and Dynamic jitter buffering can be used to offer customers better voice quality. The ITS-SP 2.0 also includes a custom application development tool that enables service providers to offer new voice over data applications in their existing networks.

MMCX 2.1 was scheduled for release on June 22, 1998. For more information about MMCX 2.1, please visit the Lucent Web site at **http://www.lucent.com/dns**. You can also contact Lucent's Multimedia Applications Customer Support center at 800-821-8204.

Northern Telecom's Webtone

Today the telecom industry is making an historic evolution toward Webtone—a shift that will enable data networks to deliver the same kind of reliability, integrity, security and capacity found in voice networks.

Nortel is meeting these challenges by striving to make next-generation Webtone networks a reality. The Internet Voice Button is its flagship product: Imagine turning on one of a variety of intelligent information devices at work, home or on the road and being within easy reach of an interactive, multidimensional world of sound, video, text and images. This is Nortel's Webtone opportunity.

A Webtone announces the immediate, instant availability of Web pages, e-mail, teleconferencing, voice conversations, text files, faxes, videos, home shopping and banking, interactive games, and every other type of digital information.

The evolution toward Webtone will change everything. By transforming how people communicate, it will transform the way businesses operate, governments define the public interest, and knowledge is created and shared. The Webtone opportunity will afford individuals the power to access a variety of media through a variety of devices. A Webtone network will have the intelligence to deliver information that will help make our lives more enjoyable, productive and effective.

Nortel is committed to building Webtone networks that carry the Internet and data traffic with the same kind of reliability, integrity, security and capacity that we take for granted in the familiar world of dialtone. Follow the links below to further explore the Webtone opportunity and learn more about Nortel's Webtone initiatives.

You should know that MICOM, also mentioned in this book for its developments and contribution in the VoIP industry, is a Nortel wholly-owned subsidiary, providing leading-edge solutions that integrate data, voice, fax and LAN communications over a wide-area link connecting company-wide locations.

Internet Voice Button

With the press of a button on a Web page, customers visiting a business Web site can call the business to place an order, request service or ask for more information—they never have to dial a number or leave their Internet session. The call won't cost them a penny.

For their convenience, customers can:

- ✔ call the business using either
  - a regular telephone, (if they have two or more telephone lines) or
  - an Internet phone, if they are using a single line to connect to the Internet
- ✔ Use popular VoIP phone clients such as Microsoft NetMeeting, supporting the H.323 protocol for Vo IP
- ✔ Use text chat to exchange additional information with the business representative.

How It Works

When a customer clicks on Voice Button on a business Web page, information is sent to the Voice Button server. This information includes the business telephone number to be called and the customer's calling preferences. These preferences are set when a customer uses Voice Button for the first time. A user-friendly configuration screen collects information including the customer's preference for using a regular phone or an Internet phone.

Voice Button initiates a call to the customer using the method selected during the configuration. When the customer answers this call, Voice Button initiates a second call to the business representative. While this call is being established, a customized announcement may be played for the customer. The business representative answers the call and is connected with the customer. A text chat facility is also supported to facilitate the exchange of information.

TIP

For more information about Voice Button, contact Nortel by e-mail at **vbutton@nortel.com**, or call them using Voice Button technology by accessing their Web site at **http://www.northerntelecom.com/**. Their telephone number is 613-765-7354 or 800-4NORTEL.

What's Next

This chapter outlined some of the main vendors actively pursuing VoIP solutions and development of products. Many already offer products and services, while others are catching up. I believe most of the big telcos will join the effort, but for now, they must concentrate on trying to convince their stockholders and boards of directors about the need completely to change the way they do business in order to adopt VoIP's technologies at full force.

There could be an endless list of vendors and suppliers of VoIp technology and products, but that would be counterproductive. The selection above is incomplete, and to update it is virtually impossible as well, as more and more companies join the VoIP market. I tried to present an overall picture of where VoIP's market is and what some of its major players are doing. Check the companies' Web sites and the latest updates of their developments.

The next chapter discusses the Real Time Streaming Protocol (RTSP) and its potential effect on VoIP, once it is fully available.

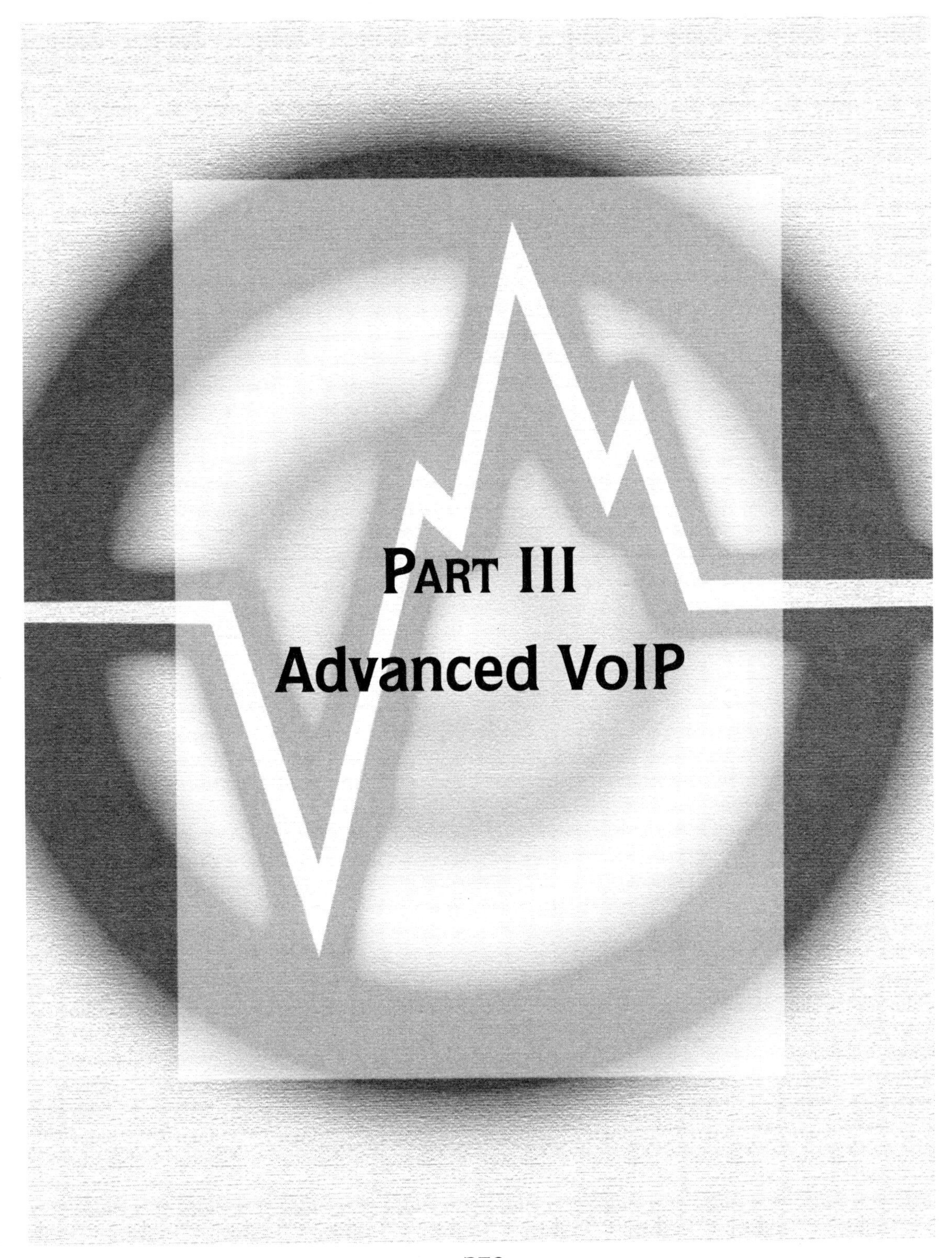

PART III
Advanced VoIP

CHAPTER 10

The Real-Time Streaming Protocol

This chapter provides a review of the Real-Time Streaming Protocol (RTSP), as described in RFC 2326 as standards track, in April 1998, proposed by H. Schulzrinne, of Columbia University, A. Rao of Netscape, and R. Lanphier of RealNetworks.[1]

1 Copyright © The Internet Society (1998). All Rights Reserved.
This document and translations of it may be copied and furnished to others, and derivative works that comment on or otherwise explain it or assist in its implementation may be prepared, copied, published and distributed, in whole or in part, without restriction of any kind, provided that the above copyright notice and this paragraph are included on all such copies and derivative works. However, this document itself may not be modified in any way, such as by removing the copyright notice or references to the Internet Society or other iNternet organizations, except as needed for the purpose of developing Internet standards in which case the procedures for copyrights defined in the Internet Standards process must be followed, or as required to translate it into languages other than English.

It is important to review RFC 2326, as it specifies the RTSP, an application-level protocol for control over the delivery of data with real-time properties. RTSP provides an extensible framework to enable controlled, on-demand delivery of real-time data, such as audio and video. Sources of data can include both live data feeds and stored clips. This protocol is intended to control multiple data delivery sessions, provide a means for choosing delivery channels such as UDP, multicast UDP and TCP, and provide a means for choosing delivery mechanisms based upon RTP (RFC 1889).

The Real-Time Streaming Protocol

The Real-Time Streaming Protocol (RTSP) establishes and controls either a single or several time-synchronized streams of continuous media such as audio and video. It does not typically deliver the continuous streams itself, although interleaving of the continuous media stream with the control stream is possible. Thus, RTSP acts as a "network remote control" for multimedia servers.

There is no notion of an RTSP connection; instead, a server maintains a session labeled by an identifier. An RTSP session is in no way tied to a transport-level connection such as a TCP connection. During an RTSP session, an RTSP client may open and close many reliable transport connections to the server to issue RTSP requests. Alternatively, it may use a connectionless transport protocol such as UDP.

The streams controlled by RTSP may use RTP, but the operation of RTSP does not depend on the transport mechanism used to carry continuous media. The protocol is intentionally similar in syntax and operation to HTTP/1.1. This enables the extension mechanisms to HTTP to be added, in most cases, to RTSP. But RTSP is not like HTTP, it differs in many ways, as outlined below:

- ✔ An RTSP server needs to maintain state by default in almost all cases, as opposed to the stateless nature of HTTP.
- ✔ Both an RTSP server and client can issue requests.
- ✔ Data is carried out-of-band by a different protocol, although there is an exception to it, which is discussed later in this chapter.
- ✔ RTSP introduces a number of new methods and has a different protocol identifier.
- ✔ RTSP is defined to use ISO 10646 (UTF-8) rather than ISO 8859-1, which is consistent with current HTML internationalization efforts.
- ✔ The Request-URI always contains the absolute URI. Also, HTTP/1.1 carries only the absolute path in the request and puts the host name in

a separate header field, so that it can continue to be backward-compatible with a historical blunder, which makes virtual hosting easier, where a single host with one IP address hosts several document trees.

RTSP supports the following operations:

- ✔ **Retrieval of media from media server**—The client can request a presentation description via HTTP or some other method. If the presentation is being multicast, the presentation description contains the multicast addresses and ports to be used for the continuous media. If the presentation is to be sent only to the client via unicast, the client provides the destination for security reasons.
- ✔ **Invitation of a media server to a conference**—A media server can be invited to join an existing conference, either to play back media into the presentation or to record all or a subset of the media in a presentation. This mode is useful for distributed teaching applications. Several parties in the conference may take turns pushing the remote control buttons.
- ✔ **Addition of media to an existing presentation**—Particularly for live presentations, it is useful if the server can tell the client about additional media becoming available.
- ✔ **Proxies, tunnels and caches** as in HTTP/1.1 may handle RTSP requests.

Properties of RTSP

RTSP is an extendable protocol, as new methods and parameters can be easily added to it. RTSP is also very easy to parse both by standard HTTP or MIME parsers. Since RTSP re-uses Web security mechanisms, it is fairly secure. All HTTP authentication mechanisms such as basic and digest authentication are directly applicable.

Another characteristic of RTSP is its transport-independence. It can use either an unreliable datagram protocol, such as UDP, a reliable datagram protocol, such as RDP, or a reliable stream protocol such as TCP as it implements application-level reliability.

RTSP is multi-server capable. Thus, each media stream within a presentation can reside on a different server. The client automatically establishes several concurrent control sessions with the different media servers. Media synchronization is performed at the transport level.

Control of recording devices can be executed with RTSP, which is able to control both recording and playback devices, as well as devices that can alternate between the two modes, such as VCRs.

Further, RTSP is suitable for professional applications, as it supports frame-level accuracy through SMPTE time stamps to allow remote digital editing. Also, it is presentation-description neutral, hence not imposing a particular presentation description or metafile format, and can convey the type of format to be used. However, the presentation description must contain at least one RTSP URI.

One of the main characteristics of this protocol is its proxy and firewall friendliness. For that, both application and transport-layer firewalls such as SOCKS should readily handle the protocol. A firewall may need to understand the setup method to open a hole for the UDP media stream.

Another major characteristic of RTSP is its HTTP-friendliness. Where sensible, RTSP reuses HTTP concepts, so that the existing infrastructure can be reused. This infrastructure includes Platform for Internet Content Selection (PICS), for associating labels with content. However, RTSP does not just add methods to HTTP since the controlling continuous media requires Server State in most cases. For additional information on RTSP characteristics, check the RFC 2326 for full information.

Next Step

Appendix A provides a list of VoIP Vendors and Appendix B provides a comprehensive glossary of terms related to VoIP.

APPENDIX A

List of Suppliers

TABLE 1 A Sampling of SONET/SDH Suppliers

Provider	Service	Availability	Description	Coverage	Bandwidth	Guarantees
Ameritech Corp. Hoffman Estates, IL 708-248-2000	Unnamed	1994	Rings and point-to-point region	90% of Ameritech	155, 622 Mbit/s; 2.4 Gbit/s	1 month free if failure is greater than 1 minute per month
AT&T Contact local sales office	Accunet T.155	1994	Point-to-point	8 cities tariffed; service available in 160 locations	155, 622 Mbit/s	Outages from 1 minute to 1 hour, 5% refund of monthly fee; 9 hours or more, 50% refund of monthly fee
Bellsouth Corp. Atlanta 404-982-7000	Smart Ring	1991	Rings	Bellsouth region	155, 622 Mbit/s	1 month credit for 2.5-second failure or greater
British Telecom-munications PLC (BT) London 44-171-932-7894	Megastream Genus	1994	Point-to-point	Trial service in Manchester and London	2 Mbit/s	Not yet determined
Colt Telecom-munications Ltd. London 44-171-390-3900	Coltlink, Coltline	London, 1992; Frankfurt, Germany, early 1996	Point-to-point	London, Frankfurt	300 bits/s to 155 Mbit/s	None
Energis Communications Ltd. London 44-171-206-5800	SDH Premium and SDH Core	1994	Rings and point-to-point	Nationwide	2, 34, 140, Mbit/s	None
LDDS Worldcom Tulsa, OK 918-494-8999	Unnamed	1994	Point-to-point	Nationwide	155, 622 Mbit/s	Case-by-case credit for outages
MFS Datanet Inc. San Jose, CA 408-975-2200	High-Speed LAN Interconnect	1992	Rings	36 U.S. cities, London, Paris, Frankfurt, Germany, and Stockholm	155 Mbit/s	None

continued on next page

Provider	Service	Availability	Description	Coverage	Bandwidth	Guarantees
Nippon Telephone and Telegraph Corp. (NTT) Tokyo 81-3-3509-8694	Super High-Speed Leased Circuit Service	1995	Rings	Nationwide	51 and 155 Mbit/s	None
Nynex Corp. White Plains, NY 914-644-7600	SONET	1995	Rings and point-to-point	Nynex region	55, 622 Mbit/s; 12.4 Gbit/s	Monthly charge refunded for 1-minute outage per month
Pacific Bell San Francisco 510-867-7258 Service	Fastrak SONET Ring and Access	1995	Rings and point-to-point	San Francisco Bay and Los Angeles areas	1.544 Mbit/s to 2.488 Gbit/s	Entire monthly service charge credited for outages of 2 hours or more hours
Sprint Corp. Kansas City, MO 703-318-7740	Clearline 155	1995	Point-to-point leased lines	36 U.S. cities	155, 622 Mbit/s; 2.4 Gbit/s	Case-by-case basis
Teleport Communications Group Staten Island, NY 718-983-2000	Omnilink Services	1994	Rings and point-to-point	40 U.S. cities	55, 622 1Mbit/s; 2.4 Gbit/s	None
US West Advanced Communications Services Denver 303-793-6500	Synchronous Service Transport (SST)	1993	Point-to-point 14-state region	US West	1.544, 45, 155, 622 Mbit/s; 1.2, 2.4 Gbit/s	Case-by-case credit

TABLE 2 ATM Around the World

Provider	Service	Availability	Coverage	Description	Bandwidth	Frame relay/ SMDS
Ameritech Corp. Hoffman Estates, IL 708-248-2000	Ameritech ATM Service	1994	Chicago, Cleveland, Columbus, Ohio Dayton, Ohio, Detroit, Milwaukee	CBR/VBR	45 and 155 Mbit/s	1996/No

continued on next page

Provider	Service	Availability	Coverage	Description	Bandwidth	Frame relay/ SMDS
AT&T Contact local sales office	AT&T Interspan	1994	Nationwide	CBR/VBR with SNMP management	45 and 155 Mbit/s	Late 1995/No
Bellsouth Corp. Atlanta 404-982-7000	Fast Packet Transport Services	1993	North Carolina	CBR/VBR	1.544-45 Mbit/s (access); 155 Mbit/s (backbone)	Yes/Yes
Deutsche **Telecom AG** Bonn 49-228-181-0	ATM Service	1995	20 cities	VBR	2, 3, 155 Mbit/s	No/Yes
France Telecom Paris 33-1-44-44-53-14	Transrel ATM	1994	11 French cities	ATM-based LAN interconnect for IP traffic only	2-25 Mbit/s	No/Planned
IBM Global **Network** Paris 33-1-41-88-60-00	IBM Business Port	1995	U.S., Canada, Europe	ATM infrastructure for multiprotocol routing, SNA, voice	8 Mbit/ s (backbone); 64-256 kbit/ s (access)	Yes/No
LDDS Worldcom Tulsa, OK 918-561-6098	Channel Networking and LAN Connection	1993	Nationwide	Mainframe to host and LAN interconnect; VBR	1.54 and 45 Mbit/s (access); 45 Mbit/s (backbone)	Late 1995/No
MCI **Communications** **Corp.** Washington, DC 202-872-1600	Hyperstream ATM Service	1994	Nationwide	CBR/VBR/UBR	45 and 155 Mbit/s (access); 155 Mbit/s (backbone)	Early 1996/ Early 1996
MFS Datanet Inc. San Jose, CA 408-975-2200	MFS Datanet ATM Service	1993	18 U.S. cities	CBR/VBR	45 and 155 Mbit/s (access); 155 Mbit/s (backbone)	Yes/No
Pacific Bell San Francisco 510-823-2558	Pacific Bell Fastrak Atm/Cell Relay Service	1993	Los Angeles, Monterey, Sacramento, San Diego, San Francisco	CBR/VBR (backbone)	45 and 155 Mbit/s (access); 155 Mbit/s	No/No

continued on next page

Provider	Service	Availability	Coverage	Description	Bandwidth	Frame relay/ SMDS
SBC Communications Corp. St. Louis 314-235-9800	Southwestern Bell ATM Cell Relay Service	1996	Metropolitan areas in Arkansas, Kansas, Missouri, Oklahoma, and Texas	CBR/VBR	45 and 155 Mbit/s	Yes/Yes
Sprint Corp. Kansas City, MO 703-318-7740	Sprint ATM Service	1993	Nationwide	CBR/VBR	1.544, 45 Mbit/s (access); 45 and 155 Mbit/s (backbone)	Late 1995/No
Stentor Alliance Ottawa 613-781-8798	Depends upon operator	1993	All major metropolitan areas in 10 provinces and territories	LAN interconnect for Ethernet and token ring; videoconferencing	10 Mbit/s (access)	N/A for both
Swiss Telecom PTT Bern 41-31-338-7393	SwissWAN	1995	Basel, Bern, Geneva, Lausanne, Lugano, Zurich	LAN interconnect; CBR/VBR	155 Mbit/s	Yes/Yes
Telecom Finland Ltd. Helsinki 358-2040-2964	Datanet ATM	1994	15 Finnish cities	LAN interconnect	64 kbit/ s to 155 Mbit/ s	Yes/No
Telia AB Stockholm 46-8-713-1975	Telia City Services	1995	Stockholm, Gothenburg, Malmo	ATM bearer services; CBR/VBR; permanent and temporary connections	155 Mbit/s	No/No
Unitel Communications Inc. Toronto 416-345-2000	Canarie Network	Late 1995	Major Canadian cities	VBR	45 Mbit/s	No/No
US West Enterprise Networking Services Denver 303-965-9286	Enterprise Networking ATM Service	1995	Oregon	ABR/VBR	45 and 155 Mbit/s (access); 155 Mbit/s (backbone)	Early 1996/ Late 1996, early 1997

ABR = Available bit rate
VBR = Variable bit rate
UBR = Unspecified bit rate
N/A = Not applicable
CBR = Constant bit rate
SVC = Switched virtual circuit
SMDS = Switched Multimegabit Data Service

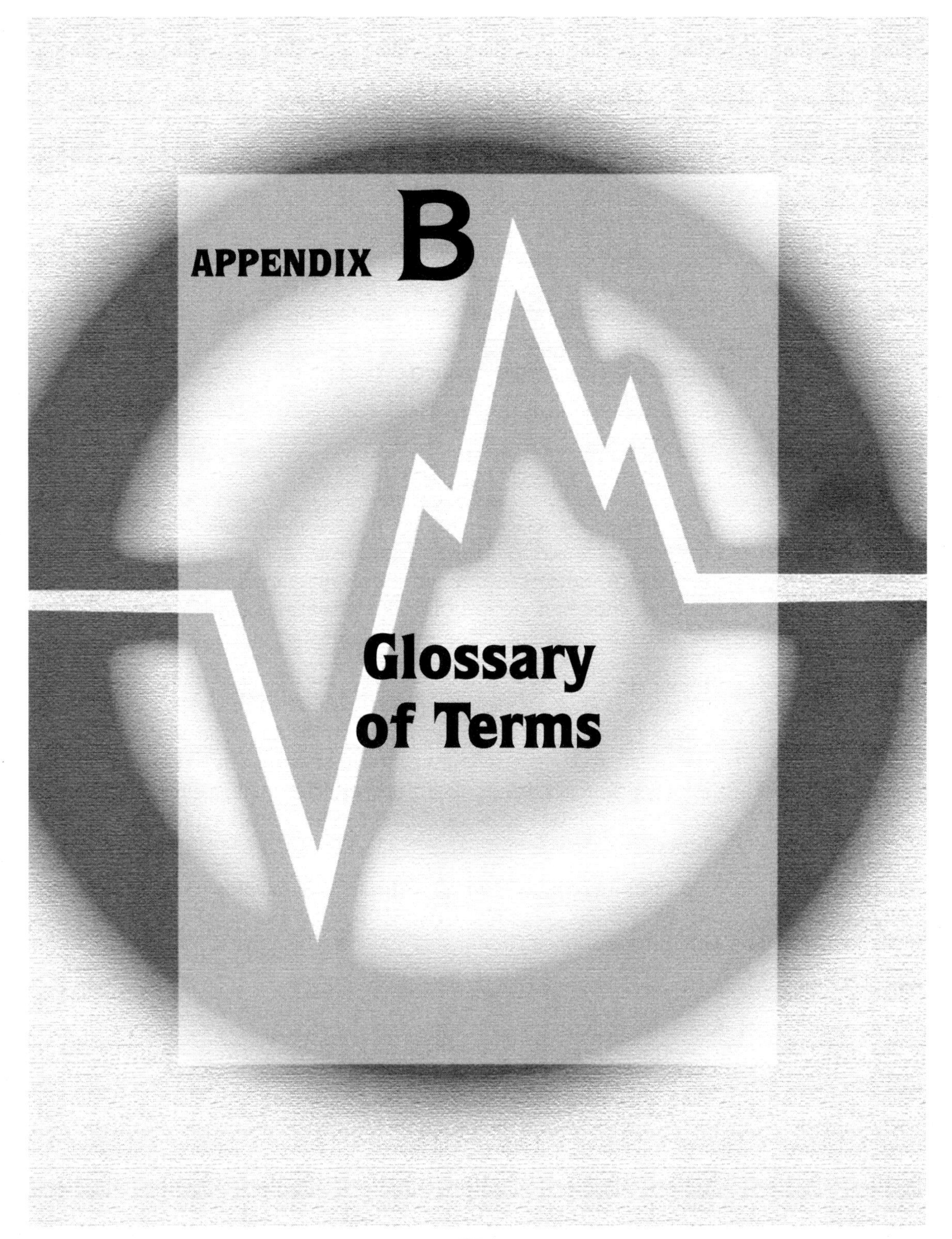

APPENDIX B

Glossary of Terms

Adjacency A relationship formed between neighboring routers for the purpose of exchanging routing information.

Aggregate control The control of the multiple streams using a single timeline by the server. For audio/video feeds, this means that the client may issue a single play or pause message to control both the audio and video feeds.

American National Standards Institute (ANSI) The principal standards development body in the United States. It consists of voluntary members that represent the US in the International Standards Organization (ISO). Membership includes manufacturers, common carriers, and other national standards organizations, such as the Institute of Electrical and Electronic Engineers (IEEE).

American Wire Gauge (AWG) A wire diameter specification. The lower the AWG number, the larger the wire diameter.

Amplitude The maximum value of varying wave forms.

Anycast, anycast address An identifier for a set of interfaces that typically belong to different nodes; a method developed for IPv6, of sending a datagram or packet to a single address with more than one interface. The packet is usually sent to the "nearest" node in a group of nodes, as determined by the routing protocols' measure of distance. Compare to **multicast** and **unicast**.

Application-Layer Gateway (ALG) In modern usage, the term application gateway refers to systems that do translation from some native format to another for example, a gateway that permits communication between TCP/IP systems and OSI systems. An application-layer gateway converts protocol data units (PDU) from one stack's application protocol to the other stack's application protocol. Application-layer gateways act as origination and termination points for communications between realms.

Application Programming Interface (API) A set of tools, routines, and protocols used as building blocks by programmers to develop programs. Using APIs helps keep applications consistent with the operating environment.

ASIC Application-Specific Integrated Circuit

Asymmetric Digital Subscriber Line (ADSL) An xDSL technology in which modems attached to twisted-pair copper wires transmit from 1.5 to 8 Mbps downstream (to the subscriber) and from 16 to 640 Kbps upstream, depending on the line distance.

Asynchronous Transfer Mode (ATM) A cell-switching and multiplexing technology that provides high-speed backbone support defined at 155 Mb/s and 622 Mb/s and having a 53-byte fixed-length cell consisting of a 5 byte header for routing information and 48 bytes of data.

Attachment Unit Interface (AUI) A 15-pin shielded, twisted pair Ethernet cable optionally used to connect network devices and a MAU.

Authentication The process of knowing that the data received is the same as the data sent and that the sender is the actual sender. This is usually verified by a password but since passwords can be guessed or discovered, a system that requires an encrypted password and a key to decrypt it are becoming popular.

Autonomous System (AS) A collection of CIDR IP address prefixes under common management. An autonomous system is a set of routers under a single technical administration. An AS uses one or more interior-gateway protocols and common metrics to route packets within the AS. An AS uses an exterior-gateway protocol to route packets to other autonomous systems. The administration of an AS appears to other ASs to have a single coherent interior routing plan and presents a consistent picture of what networks are reachable.

Backbone network The major transmission path for network interconnection.

Bandwidth The signaling rate of a LAN or WAN circuit or the number of bits or bytes that can be transmitted over the channel each second and measured by electrical engineers in hertz (Hz). *See also* **latency**.

BNC Connector Bayonet Neill-Concelman connector: a type of connector used for attaching coax cable to electronic equipment; it can be attached or detached more quickly than screw-type connectors. ThinWire Ethernet (IEEE 802.3 10BASE2) uses BNC connectors.

Broadband A data transmission technique allowing multiple high-speed signals to share the bandwidth of a single cable via frequency division multiplexing.

Broadcast A type of data communication where a source sends one copy of a message to all the nodes on the network even if any node does not want to receive such messages. *See also* **anycast**, **unicast**, **multicast**, and **IP multicasting**.

Broadcast domain The part of a network that receives the same broadcasts.

Broadcast network A network that supports more than two attached routers, and has the capability to address a single physical message to all of the attached routers.

Building backbone subsystem This is the link between the building and campus backbone.

Campus backbone subsystem This provides the link between buildings and contains the cabling and cross-connects between clusters of buildings within a site.

Carrier Sense Multiple Access with Collision Detection (CSMD/CD) The channel access method used by the Ethernet and ISO 8802-3 LANs. Each station waits for an idle channel before transmitting and detects overlapping transmissions by other stations.

Carrierless Amplitude Phase (CAP) Modulation This version of quadrature amplitude modulation (QAM) stores parts of a modulated message signal in memory and then reassembles the parts in the modulated wave. The carrier signal is suppressed before transmission because it contains no information and is reassembled at the receiving modem (hence the word "carrierless" in CAP).

Central Office (CO) A facility that contains the lowest node in the hierarchy of switches that comprises the public telephone network.

Channel The data path between two nodes.

Class A IP address A type of unicast IP address that segments the address space into many network addresses and few host addresses.

Class B IP address A type of unicast IP address that segments the address space into a medium number of network and host addresses.

Class C IP address A type of unicast IP address that segments the address space into many host addresses and few network addresses.

Class D IP addresses This specifies multicast host groups in IPv4 based networks. The Internet standard in "dotted decimal" notation assigns this host group addresses ranging from 224.0.0.0 to 239.255.255.255.

Client The client requests continuous media data from the media server.

Client/server architecture A network architecture in which the protocols in use govern the behavior of workstations so each one works as either a client or a server. Users run applications on client machines while server machines manage network resources.

CLNP An OSI connectionless network protocol.

Conference A multiparty, multimedia presentation, where "multi" implies greater than or equal to one.

Connection A transport-layer virtual circuit established between two programs for the purpose of communication.

Connectionless protocol A type of network protocol where a host can send a message without establishing a connection with the recipient. The host puts the message onto the network, provides the destination address and hopes that the message arrives at its destination.

Connection-oriented protocol A protocol that requires the establishment of a channel between the sender and receiver before transmitting any data. The telephone, TCP, and HyperText Transmission Protocol (HTTP) are all examples of connection-oriented protocols.

Consumer Defined in the Multicast Transport Protocol as a transport capable only of receiving user data. It can transmit control packets, such as negative acknowledgements, but can never transmit requests for the transmit token or any form of data or empty messages.

Container file A file that may contain multiple media streams, which often comprise a presentation when played together. RTSP servers may offer aggregate control on these files, though the concept of a container file is not embedded in the protocol.

Continuous media Data where there is a timing relationship between source and sink; that is, the sink must reproduce the timing relationship that existed at the source. The most common examples of continuous media are audio and motion video. Continuous media can be real-time (interactive), where there is a "tight" timing relationship between source and sink, or streaming (playback), where the relationship is less strict.

Convergence The amount of time it takes for a change to a routing topology to propagate throughout the network.

Core-Based Trees (CBT) Routing Protocol The CBT routing protocol is characterized by a single tree shared by all members of the group. Group members receive multicast traffic over this shared tree regardless of the source of the message. A small number of core routers constructs the tree, and routers can join the tree by sending a join message to the core.

Core network A combination of switching offices and transmission plant that connects switching offices. In the US local exchange, core networks are linked by several competing interexchange networks. In the rest of the world, core networks extend to national boundaries.

Counter-rotating ring A method of using two ring networks going in opposite directions (such as in FDDI) to provide redundancy. The network interfaces can change the path of the ring the data flows around. This preserves the ring and the operation of the LAN even if some of the cable is unplugged or cut, or if a device on the ring fails in such a way that it cannot transmit data around the ring.

CPE Customer Premises Equipment

Data Circuit Terminating Equipment (DCE) An interface typically found in modems or similar devices that provide clocking as well as switching services between DTE.

Data Link Connection Identifier (DLCI) A 10-bit value included in the address field of a frame relay packet that uniquely identifies each virtual circuit at each frame relay.

Data Terminal Equipment (DTE) An interface typically embodied in computers, terminals, or routers that act as terminating equipment for a given network.

Datagram Term used in IPv4; the format for a packet of data sent on the Internet to a specific destination address. It specifies standards for the header information. In IPv6, datagrams are known as packets.

Dense-mode multicast routing protocols A category of routing protocol that assumes that multicast group members are densely distributed throughout the network. The basic assumption is that almost all the hosts on the network belong to the group. Dense-mode routing protocols included the Distance Vector Multicast Routing Protocol (DVMRP), Multicast Open Shortest Path First (MOSPF), and Protocol-Independent MulticastDense Mode (PIM-DM) protocols. *See also* **sparse-mode routing protocols**.

Dial up A type of communications established by a switched-circuit connection using the public telephone network.

Digital Loop Carrier (DLC) The carrier's local loop infrastructure that connects end users located more than 18,000 feet or 3.5 miles away from the central office. DLC systems consist of physical pedestals containing line cards that concentrate residential links onto digital circuits.

Digital Subscriber Line (DSL) A local loop access technology that calls for modems on either end of copper twisted-pair wire to deliver data, voice, and video information over a dedicated digital network.

Digital Subscriber Line Access Multiplexer (DSLAM) Multiplexing equipment that contains a high concentration of central office splitters, xDSL modems, and other electronics to connect traffic to the WAN.

Discrete Multi-tone (DMT) Modulation A wave modulation scheme that discretely divides the available frequencies into 256 sub-channels or tones to avoid high-frequency signal loss caused by noise on copper lines.

Discrete Wavelet Multi-Tone (DWMT) A variant of DMT modulation. DWMT goes a step further in complexity and performance by creating even more isolation between sub-channels.

Distance Vector Multicast Routing Protocol (DVMRP) The first protocol developed to support multicast routing and used widely on the Mbone. RFC 1075 describes this protocol. DVMRP assumes that every host on the network belongs to the multicast group. Multicast messages pass over all router interfaces as they cross the network and form a spanning tree to all members of the multicast group. DVMRP uses a distance-vector protocol such as the Routing Information Protocol (RIP) to maintain a current image of the network topology. Both DVMRP and RIP use the number of hops in the path as the distance metric.

Domain Name Service (DNS) The name service of the TCP/IP protocol family, which provides information about computers on local and remote networks. DNS is an Internet-wide hierarchical database.

DPP Distributed Packet Pipelining

DTE/DCE The interface between data terminal equipment (DTE) and data circuit-terminating equipment (DCE); one of the most common in networking.

E1 The European basic multiplex rate that carries 30 voice channels in a 256-bit frame transmitted at 2.048 Mbps.

Echo cancellation A technique used by ADSL, V.32, and V.34 modems that isolates and filters unwanted signal energy from echoes caused by the main transmitted signal.

Entity The information transferred as the payload of a request or response. An entity consists of metainformation in the form of entity-header fields and content in the form of an entity.

European Telecom Standards Institute (ETSI) A consortium of manufacturers, service carriers, and others responsible for setting technical standards in the European telecommunications industry.

Fanout The degree of replication in a multicast tree or the number of copies of a call in a switch, associated with IP multicasting and ATM.

Fast Ethernet A nickname for the 100Mbps version of IEEE 802.3

Fast packet A data transmission technique where the packet is transmitted without any error checking at points along the route. The end-points have the responsibility of performing any error checking.

Fiber Distributed Data Interface (FDDI) A set of ANSI/ISO standards that define a high-bandwidth (100 Mbps) general purpose LAN. FDDI primarily runs over optical fiber but can run over copper. FDDI provides synchronous and asynchronous services between computers and peripheral equipment in a time-token passing dual ring configuration.

FIRE Flexible Intelligent Routing Engine

Flow-based A proprietary implementation of Layer 3 switching that investigates only the first packet of data, switching the remaining packets at Layer 2

Fragment A portion of a packet/frame and often a part of an Ethernet frame left over from a collision. In IP terminology, fragment means a packet that is the result of splitting a larger packet into smaller ones.

Frame In telecommunications, a unit of data that is transmitted between network points complete with addressing and necessary protocol control information.

Frame relay network A network consisting of frame relay switches, offering a bare-bones link-layer service for fast bulk packet transmission.

Frequency The rate of signal oscillation in hertz (Hz).

Frequency Division Multiplexing (FDM) A technique that divides the available bandwidth of a channel into a number of separate channels.

Full-duplex The property of a data-communications line that provides Independent, simultaneous two-way transmission in both directions, as opposed to half-duplex transmission. The alternatives are half-duplex and simplex.

Gateway An intermediate destination by which packets are delivered to their ultimate destination; a host address of another router that is directly reachable through an attached network. As with any host address it can be specified symbolically.

Geostationary Satellite (GSAT) A satellite that orbits the earth directly over the equator, approximately 22,000 miles up. A complete rotation around the earth takes 24 hours. Weather satellites are examples of this type.

Gigabit Ethernet High-speed version of Ethernet (a billion bits per second) under development by the IEEE.

Half-duplex A possible property of a data-communications line: that data can be transferred in either direction, but only in one direction at a time. If the line is sufficiently high-speed, then to a human, it may appear that data transfer is simultaneous in both directions if the two ends quickly take turns transferring. The alternatives are full-duplex and simplex.

Heartbeat An interval of time nominally measured in milliseconds and a key parameter in the transport's state. It can be adapted to the requirements of the transport's client to provide the desired quality of service. Also Ethernet defined SQE signal quality test function.

Hertz A frequency unit equal to one cycle per second.

High bit-rate Digital Subscriber Line (HDSL) An xDSL technology in which modems on either end of two or more twisted-pair lines deliver symmetric T1 or E1 speeds. Currently, T1 requires two lines and E1 requires three.

High-Definition TeleVision (HDTV) A system of transmitting television signals at 24 Mbps, which increases the horizontal lines of resolution from 480 to 560 lines per display.

Host Group All hosts belonging to a multicast session. The membership of a host group is dynamic: hosts can join and leave the group at any time. There can be any number of members in a host group and the members can be located anywhere on the local network or on the Internet. A host can be a member of more than one group at a time.

HTTP HyperText Transfer Protocol

ICMP destination unreachable indication An error indication returned to the original sender of a packet that cannot be delivered for reasons outlined in ICMP protocol. If the error occurs on a node other than the node originating the packet, an ICMP error message is generated. If the error occurs on the originating node, an implementation is not required actually to create and send an ICMP error packet to the source, as long as the upper-layer sender is notified through an appropriate mechanism, for example, the return value from a procedure call. Note, however, that an implementation may in some cases find it convenient to return errors to the sender by taking the offending packet, generating an ICMP error message, and then delivering it locally through generic error-handling routines.

IEEE Institute of Electrical and Electronic Engineers

Integrated Services Digital Network (ISDN) All digital service provided by telephone companies. Provides 144 Kbps over a single phone line (divided in two 64 Kbps "B" channels and one 16 Kbps "D" channel).

InterExchange Carrier (IEC) A long-distance service provider.

Interface A system's attachment point to a link. It is possible for a system to have more than one interface to the same link. Interfaces are uniquely identified by IP unicast addresses; a single interface may have more than one such address. An interface can be a connection between a router and one of its attached networks. A single IP address, domain name, or interface name can specify a physical interface (unless the network is an unnumbered point-to-point network).

Internet Assigned Numbers Authority (IANA) The central coordinator for the assignment of unique parameter values for Internet protocols. The Internet Society (ISOC) and the Federal Network Council (FNC) charter the IANA to act as the clearinghouse to assign and coordinate the use of numerous Internet protocol parameters.

Internet datagram The unit of data exchanged between an Internet module and the higher level protocol together with the internet header.

Internet Engineering Task Force (IETF) An international group of network designers, operators, vendors, and researchers, closely aligned to the Internet Architecture Board and chartered to work on the design and engineering of TCP/IP and the global Internet. The IETF is divided into groups or areas, each with a manager and is open to any interested individual.

Internet Group Management Protocol (IGMP) Multicast routers use this protocol to learn the existence of host group members on their directly attached subnets. IP hosts use IGMP to report their host group memberships to any immediately neighboring multicast routers. IGMP messages are encapsulated in IP datagrams, with an IP protocol number of 2. RFC1112 describes IGMP, which is considered as an extension to ICMP and occupies the same place in the IP protocol stack.

Internet Protocol (IP) The protocol or standard at the network level of the Internet that defines the packets of information and routing them to remote nodes, and the method of addressing remote computers and routing packets to remote hosts.

Internet Service Provider (ISP) A business that provides subscription services, such as online information retrieval software, bulletin boards, electronic mail, and so on to users for a fee. ISPs are domains under the control of a single administration that share their resources with other domains.

Internetwork Packet Exchange protocol (IPX) A datagram protocol found in Novell NetWare networks. It is similar to UDP and together with SPX provides connectionless services similar to UDP/IP.

InterNIC A collaborative project between AT&T and Network Solutions, Inc. (NSI) supported by the National Science Foundation. The project currently offers four services to users of the Internet.

IP Multicast A one-to-many transmission described in RFC 1112. The RFC describes IP multicasting as: "the transmission of an IP datagram to a 'host group', a set of zero or more hosts identified by a single IP destination address. A multicast datagram is delivered to all members of its destination host group with the same 'best-efforts' reliability as regular unicast IP datagrams. The membership of a host group is dynamic; that is, hosts may join and leave groups at any time. There is no restriction on the location or number of members in a host group. A host may be a member of more than one group at a time."

IP Multicast Datagram A datagram delivered to all members of the multicast host group. Such datagrams are delivered with the same best-efforts reliability as regular unicast IP datagrams.

IP Multicast Router A router supporting IGMP and one or more of the multicast routing protocols, including Distance DVMRP, MOSPF, PIM-DM, CBT, and PIM-SM.

IPX See Internetwork Packet Exchange Protocol

ISE Intelligent Switching Engine

ISO International Standardization for Organization, a special agency of the United Nations that is charged with the development of communication standards for computers. Membership in the ISO consists of representatives from international standards organizations throughout the world.

ISP Internet Service Provider

ITU International Telecommunication Union

Kbps Kilobits per second

LAN A local area network; a communication network that spans a limited geographical area. LANs can differ from one another in topology or arrangement of devices on the network, the protocols they use, and the media, such as twisted-pair wire, coaxial cables, or fiber optic cables used to connect the devices on the network.

Latency The transmission delay of the network or the minimum amount of time it takes for any one of those bits or bytes to travel across the network. *See also* **bandwidth**.

LEO system A low-earth-orbit satellite system consisting of a number of small satellites orbiting in a circular orbit at over, or nearly over, the geographic poles and flying at an altitude of a few hundred miles. Wireless access to the Internet is dependent upon this type of satellite.

LLC Logical Link Control

Local loop The line from a subscriber to the telephone company central office.

Logical link A temporary connection between source and destination nodes, or between two processes on the same node.

Logical Link Control (LLC) Part of the data-link layer of the OSI model and the link-layer control specification for the IEEE 802.x series of standards. It defines the services for the transmission of data between two stations with no intermediate switching stations. There are three versions; LLC1 is connectionless, LLC2 is connection-oriented, and LLC3 is connectionless with acknowledgment.

Logical topologies The view of the network as seen by the network's components, access methods or rules of operation.

MAC address The unique media access control 6-byte address that is associated with the network adapter card and identifies the machine on a particular network. A MAC address is also known as an Ethernet address, hardware address, station address, or physical address.

Mask A means of subdividing networks using address modification. A mask is a dotted quad specifying which bits of the destination are significant. Except when used in a route filter, Gate D only supports contiguous masks.

Maximum Transmission Units (MTU) The largest amount of data that can be transferred across a network; size is determined by the network hardware.

Mbone A virtual multicast backbone network layered on top of the physical Internet. In existence for about five years, the Mbone supports routing of IP multicast packets.

Mbps Megabits per second

Media initialization Data type/codec specific initialization. This includes such things as clock rates, color tables, etc. Any transport-independent information required by a client for playback of a media stream that occurs in the media initialization phase of stream setup.

Media parameter Parameter specific to a media type that may be changed before or during stream playback.

Media server The server providing playback or recording services for one or more media streams. Different media streams within a presentation may originate from different media servers. A media server may reside on the same or a different host as the Web server from which the presentation is invoked.

Media server indirection Redirection of a media client to a different media server.

Media stream A single media instance, e.g., an audio stream or a video stream as well as a single whiteboard or shared application group. When using RTP, a stream consists of all RTP and RTCP packets created by a source within an RTP session. This is equivalent to the definition of a DSM-CC stream.

Medium Attachment Unit (MAU) A device used to convert signals from one Ethernet medium to another.

MIB Management Information Base

Midband A communication channel with a bandwidth range of 28.8 to 56 Kbps.

Modem Contraction for modulator/demodulator. A modem converts the serial digital data from a transmitting device into a form suitable for transmission over the analog telephone channel.

Modulation The process in which the characteristics of one wave or signal are varied in accordance with another wave or signal. Modulation can alter frequency, phase, or amplitude characteristics.

Multiaccess network A physical network that supports the attachment of more than two routers. Each pair of routers on such a network can communicate directly.

Multicast Method of transmitting messages from a host using a singe transmission to a selected subset of all the hosts that can receive the messages; also a message that is sent out to multiple devices on the network by a host. *See also* **anycast**, **unicast**, **broadcast**, and **IP multicasting**.

Multicast group A group set up to receive messages from a source. These groups can be established based on frame relay or IP in the TCP/IP protocol suite, as well as in other networks.

Multicast interface An interface to a to a link over which IP multicast or IP broadcast service is supported.

Multicast link A link over which IP multicast or IP broadcast service is supported. This includes broadcast media such as LANs and satellite channels, single point-to-point links, and some store-and-forward networks such as SMDS networks.

Multicast Open Shortest Path First (MOSPF) RFC 1584 defines MOSPF as an extension to the OSPF link-state unicast routing protocol that provides the ability to route IP multicast traffic. Some portions of the Mbone support MOSPF. MOSPF uses the OSPF link-state metric to determine the least-cost path and calculates a spanning tree for routing multicast traffic with the multicast source at the root and the group members as leaves.

Multicast Transport Protocol (MTP) This protocol gives application programs guarantees of reliability. The MTP protocol could be useful when developing some types of applications, for example with distributed databases that need to be certain that all members of a multicast group agree on which packets have been received.

Multimode Fiber A type of fiber mostly used for short distances such as those found in a campus LAN. It can carry 100 Mbs/sec for typical campus distances, the actual maximum speed (given the right electronics) depending upon the actual distance. It is easier to connect to than single-mode fiber, but its limit on speed x distance is lower.

Multiplex Combining signals of multiple channels into one channel. This process provides multiple users with access to a single conductor or medium by transmitting in multiple distinct frequency bands (frequency division multiplexing, or FDM) or by assigning the same channel to different users at different times (time division multiplexing, or TDM).

Multiplexer A device that allows several users to share a single circuit and funnels different data streams into a single stream. At the other end of the communications link, another multiplexer reverses the process by splitting the data stream back into the original streams.

Multiplexing A repeater, either standalone or connected to standard Ethernet cable, for interconnecting up to eight thin-wire Ethernet segments.

Narrowband A communication channel with a bandwidth of less than 28.8 Kbps.

NetBEUI A enhanced version of the NetBIOS protocol used by Windows-based operating systems such as Windows 95 and Windows NT.

NetBIOS The Network Basic Input Output System, an API that has special functions for local-area networks and is used with the DOS BIOS.

Network Access Point (NAP) An Internet hub where national and international ISPs connect with one another. A NAP router has to know about every network on the Internet

Network address *See* **IP address**

Network Service Access Point (NSAP) The network address, or the node address of the machine, where a service is available.

Network Information Center (NIC) Central organization of a network with the authority to create network names and addresses. NIC.DDN.MIL is the specific Internet NIC that holds the authority to create root servers.

Network Information Service (NIS) Formerly known as Sun Yellow Pages, NIS is used for the administration of network-wide databases. NIS has two services, one for finding a NIS server, the other for access to the NIS databases. NIS permits dynamic updates of the database files. NIS is a non-hierarchical, replicated database which is the property of Sun Microsystems.

Node Any intelligent device connected to the network. This includes terminal servers, host computers, and any other devices (such as printers and terminals) directly connected to the network. A node can be thought of as any device that has a hardware address.

NSP Network Service Provider

Open System Interconnection (OSI) The title for a set of layered standards developed by the ISO to allow communication between the computer systems of different vendors.

OSPF Open Shortest Path First

Packet A package of data with a header that may or may not be logically complete. A series of bits containing data and control information, including source and destination node addresses, formatted for transmission from one node to another. A packet is more often a physical packaging than a logical packaging of data.

Packet-by-packet An implementation of Layer 3 switching that uses industry-wide, standard routing protocols to examine all packets and forward them to their destination entirely in Layer 3.

Participant Member of a conference; a participant may be a machine, e.g., a media record or playback server.

Peer-to-peer architecture The arrangement of communication functions and services in layers so that data transmission between logical groups or layers in a network architecture occurs between entities in the same layer of the model. With a peer-to-peer architecture all workstations in this type of network have the equivalent capabilities. *See also* **client/server architecture**.

Permanent Virtual Circuit (PVCP) A permanent logical connection set up with packet data networks such as frame relay.

Phase modulation A technique that changes the characteristics of a generated sine wave or signal so that it will carry information.

Physical layer The physical channel implements Layer 1, the bottom layer of the OSI model. The physical layer insulates Layer 2 (the data-link layer) from medium-dependent physical characteristics such as baseband, broadband or fiber-optic transmission. Layer 1 defines the protocols that govern transmission of media and signals.

Physical topologies These define the arrangement of devices and the layout of the wiring.

PIM Dense Mode (PIM-DM) Routing Protocol Protocol Independent Multicast-Dense Mode is a protocol that operates in an environment where group members are relatively densely packed. PIM-DM is similar to DVMRP in that it employs the Reverse Path Multicasting (RPM) algorithm. PIM-DM controls message processing and data packet forwarding and is integrated with PIM-SM operation so that a single router can run different modes for different groups.

PIM-Sparse Mode (PIM-SM) Routing Protocol PIM-SM is a protocol optimized for environments where group members are distributed across many regions of the Internet. To receive multicast traffic addressed to the group, routers with directly attached or downstream members are required to join a sparse-mode distribution tree by transmitting explicit join messages. To eliminate potential scaling issues, PIM-SM limits multicast traffic so that only those routers interested in receiving traffic for a particular group see it.

Point-to-point network A network joining a single pair of routers; for example, a 56Kb serial-line network.

Point-to-Point Protocol (PPP) The successor to SLIP, PPP provides router-to-router and host-to-network connections over both synchronous and asynchronous circuits.

POTS Plain Old Telephone Service

POTS splitter A passive filter that separates voice traffic from data traffic.

Presentation A set of one or more streams presented to the client as a complete media feed, using a presentation description. In most cases in the RTSP context, this implies aggregate control of those streams.

Presentation description A presentation description contains information about one or more media streams within a presentation, such as the set of encodings, network addresses and information about the content.

Protocol The set of rules to send and receive data and govern activities within a specific layer of the network architecture model. Protocols regulate the transfer of data between layers and across links to other devices and define procedures for handling lost or damaged transmissions or packets. Protocols also determine whether the network uses peer-to-peer or client/server architecture.

Protocol-Independent Multicast (PIM) Routing Protocol Developed by an IETF working group, PIM provides a standard multicast routing protocol that supports scalable inter-domain multicast routing across the Internet independent of the mechanisms provided by any particular unicast routing protocol. PIM has two modes, dense and sparse.

Public Switched Telephone Network (PSTN) A telephone system through which users can be connected by dialing specific telephone numbers.

QoS Quality of Service

Quadrature Amplitude Modulation (QAM) A bandwidth conservation process routinely used in modems, QAM enables two digital carrier signals to occupy the same transmission band width.

Random delay The random amount of time a transmission is delayed to prevent multiple nodes from transmitting at exactly the same time, or to prevent long-range periodic transmissions from synchronizing with each other.

RAP Roving Analysis Port

Rate-Adaptive Digital Subscriber Line (R-ADSL) An emerging variation of CAP it divides the transmission spectrum into discrete subchannels and adjusts each signal transmission according to line quality.

Reachability Whether or not the one-way forward path to a neighbor is functioning properly. For neighboring routers, reachability means that packets sent by a node's IP layer are delivered to the router's IP layer, and the router is indeed forwarding packets. This means the node is configured as a router, not a host. For hosts, reachability means that packets sent by a node's IP layer are delivered to the neighbor host's IP layer.

Real-Time Streaming Protocol (RTSP) This application-level protocol provides control for the delivery of data with real-time properties. RTSP enables controlled on-demand delivery of real-time data, such as audio and video.

Real-Time Transport Protocol (RTTP) RTTP provides end-to-end network transport functions for applications that transmit real-time data over multicast or unicast network services. Such applications can include audio, video, or simulation data applications.

RED Random Early Detection

Regional Bell Operating Company (RBOC) A telecommunication company formed as a result of the divestiture of AT&T. RBOCs oversee Bell operating companies.

Relay A device that interconnects LANs, different kinds of relays include repeaters, bridges, routers, and gateways.

Reliable multicast protocols Reliable multicast protocols provide for reliable transmission of datagrams from a single-source host to members of a multicast group. An example of a reliable multicast protocol is the Multicast Transport Protocol (MTP), which gives application programs guarantees of reliability. *See also* **Multicast Transport Protocol (MTP)**.

Request An RTSP request. If an HTTP request is meant, that is indicated explicitly.

Request for Comment (RFC) An official document used by the IETF to create standards for use in the Internet.

ReSerVation Protocol (RSVP) A method developed by the IETF to assist in providing QoS characteristics to communications over an IP network. The name refers to the fact that it allows the end-stations to reserve bandwidth on the network. This protocol supports requests for a specific QoS from the network for particular data streams or flows.

Response An RTSP response. If an HTTP response is meant, that is explicitly indicated.

Retention Defined in the Multicast Transport Protocol as one of the three fundamental parameters that make up the transport's state (along with heartbeat and window). Retention is a number of heartbeats, and though applied in several different circumstances, is primarily used as the number of heartbeats a producing client must maintain buffered data should they need to be retransmitted.

Reverse Address Resolution Protocol (RARP) An Internet protocol that can be used by diskless hosts to find their Internet address. *See* RFC 903.

RIP (Routing Information Protocol) An early BSD UNIX routing protocol that has become an industry standard.

RISC Reduced Instruction Set Computing

RMON Remote Monitoring

Router A device that connects two networks at the network layer (Layer 3) of the OSI model; operated like a bridge but also can choose routes through a network

Routing In networking, routing is the process of moving a packet of data from source to destination. A dedicated device called a router usually performs routing. Routing is a key feature of the Internet and enables messages to pass from one computer to another and eventually reach the target machine. Each intermediary computer performs routing by passing along the message to the next computer. Part of this process involves analyzing a routing table to determine the best path.

RSVP Resource ReserVation Protocol

RTSP session A complete RTSP "transaction", e.g., the viewing of a movie. A session typically consists of a client setting up a transport mechanism for the continuous media stream, starting the stream with Play or Record, and closing the stream with Teardown.

Sequenced Packet Exchange (SPX) A connection-oriented protocol found in Novell NetWare networks. This transport-layer protocol is similar to TCP and together with IPX provides connection services similar to TCP/IP.

Shared Ethernet An Ethernet configuration in which a number of segments are bound together in a single collision domain; hubs produce this type of configuration where only one node can transmit at a time.

Simple Network Management Protocol (SNMP) Allows a TCP/IP host running an SNMP application to query other nodes for network-related statistics and error conditions. The other hosts, which provide SNMP agents, respond to these queries and allow a single host to gather network statistics from many other network nodes.

Single-mode fiber A type of fiber optic cable used for longer distances and higher speeds such as longdistance telephone lines. *See also* **multimode fiber**.

Single-Line Digital Subscriber Line (SDSL) SDSL is HDSL over a single twisted pair.

SLA Service Level Agreement

SNAP Subnetwork Access Protocol

SONET (Synchronous Optical Network) A Bellcore-developed set of standard fiber optic-based serial standards planned for use with ATM in North America. Different types of SONET run at different speeds, use different types of fiber, and operate over different distances. There are both single-mode and multimode fiber versions.

Spanning tree An algorithm used to create a logical topology that connects all network segments, and ensures that only one path exists between any two nodes. A spanning tree is loop-free and is a subset of a network. Multicast routers construct a spanning tree from the multicast source located at the root of the tree to all the members of the multicast group.

Sparse-mode multicast routing protocols Sparse-mode routing protocols assume that the multicast group members are sparsely distributed throughout the network. The multicast group members could be distributed across many regions of the Internet. There can be just as many multicast group members in sparse-mode routing as there can be in dense-mode routing. Sparse-mode routing protocols include the Core Based Trees (CBT) and Protocol-Independent Multicast-Sparse Mode (PIM-SM) protocols. *See also* **dense-mode multicast routing protocols**.

Stream-oriented protocol A type of protocol where data is organized as a stream of bytes and uses a technique for transferring data such that it can be processed as a steady and continuous stream. With streaming, a client can start displaying the data before the entire file has been transmitted. If a client receives the data more quickly than required, this saves the excess data in a buffer. If the data do not come quickly enough, however, the presentation of the data is not smooth.

Switch A device that connects multiple network segments at the data-link layer (Layer 2) of the OSI model, it operates more simply and at higher speeds than does a router.

Switched Multimegabit Data Service (SMDS) High speed, connectionless, packet-switched, WAN networking technology.

T1 A 1.544 Mbps line; it is the same as DS1.

Telco American jargon for telephone company.

Time Division Multiplexing (TDM) A digital transmission method that combines signals from multiple sources on a common path. This common path is divided into a number of time slots and each signal or channel is assigned its own intermittent time slot, allowing the path to be shared by multiple channels.

Timed-token protocol The rules defining how the target-token rotation time is set, the length of time a station can hold the token, and how the ring is initialized.

Token A bit pattern consisting of a unique symbol sequence that circulates around the ring following frame transmission. The token grants stations the right to transmit.

Token passing A method where each station, in turn, receives and passes on the right to use the channel. In FDDI, the stations are configured in a logical ring.

Token ring Developed by IBM, this 4- or 16-Mbps network uses a ring topology and a token-passing access method.

Transmission Control Protocol (TCP) The protocol at the Internet's transport layer that governs the transmission of datagrams or packets by providing reliable, full-duplex, stream service to application protocols, especially IP. It provides reliable connection-oriented service by requiring that the sender and receiver exchange control information, or establish a connection before transmission can occur. *Contrast to* **User Datagram Protocol**.

Transport Service Access Point (TSAP) The address that uniquely defines a particular instantiation of a service and is formed by logically concatenating the node's NSAP with a transport identifier and sometimes a packet/protocol type.

Tunneling The practice of encapsulating a message from one protocol in another protocol and using the second protocol to transverse a number of network hops. At the destination, the encapsulation is stripped off and the original message is reintroduced to the network.

Twisted pair Telephone system cabling that consists of copper wires loosely twisted around each other to help cancel out any induced noise in balanced circuits.

UDP User Datagram Protocol

Unicast The method of sending a packet or datagram to a single address. This type of point-to-point transmission requires the source to send an individual copy of a message to each requester. *See also* **anycast**, **multicast**, **broadcast**, and **IP multicasting**.

Universal Coordinated Time (UCT) The number of seconds since 00:00 01/01/1970 Greenwich Mean Time.

UTP Unshielded twisted pair: one or more cable pairs surrounded by insulation. UTP is commonly used as telephone wire.

Variable MTU A link that does not have a well-defined MTU, such as an IEEE 802.5 token ring link. Many links, for example Ethernet links, have a standard MTU defined by the link-layer protocol or by the specific document describing how to run IP over the link layer.

Very high bit-rate Digital Subscriber Line (VDSL) A technology in which modems enable access and communications over twisted-pair lines at a data rate from 1.54 Mbps to 52 Mbps. VDSL has a maximum operating range from 1,000 feet to 4,500 feet on 24-gauge wire.

VLAN Virtual Local Area Network

Wide Area Network (WAN) A geographically dispersed network.

Window One of the fundamental elements of the transport's state that can be controlled to affect the QoS provided to the client. It represents the number of user-data carrying packets that may be multicast into the Web during a heartbeat by a single member.

XDSL The X represents the various forms of digital subscriber line (DSL) technologies: ADSL, R-ADSL, HDSL, SDSL, or VDSL.

Index

B

C

G

H

I

J

L

M

N

O

P

Q

R

S

SOFTWARE AND INFORMATION LICENSE

The software and information on this diskette (collectively referred to as the "Product") are the property of The McGraw-Hill Companies, Inc. ("McGraw-Hill") and are protected by both United States copyright law and international copyright treaty provision. You must treat this Product just like a book, except that you may copy it into a computer to be used and you may make archival copies of the Products for the sole purpose of backing up our software and protecting your investment from loss.

By saying "just like a book," McGraw-Hill means, for example, that the Product may be used by any number of people and may be freely moved from one computer location to another, so long as there is no possibility of the Product (or any part of the Product) being used at one location or on one computer while it is being used at another. Just as a book cannot be read by two different people in two different places at the same time, neither can the Product be used by two different people in two different places at the same time (unless, of course, McGraw-Hill's rights are being violated).

McGraw-Hill reserves the right to alter or modify the contents of the Product at any time.

This agreement is effective until terminated. The Agreement will terminate automatically without notice if you fail to comply with any provisions of this Agreement. In the event of termination by reason of your breach, you will destroy or erase all copies of the Product installed on any computer system or made for backup purposes and shall expunge the Product from your data storage facilities.

LIMITED WARRANTY

McGraw-Hill warrants the physical diskette(s) enclosed herein to be free of defects in materials and workmanship for a period of sixty days from the purchase date. If McGraw-Hill receives written notification within the warranty period of defects in material or workmanship, and such notification is determined by McGraw-Hill to be correct, McGraw-Hill will replace the defective diskette(s). Send request to:

Customer Service
McGraw-Hill
Gahanna Industrial Park
860 Taylor Station Road
Blacklick, OH 43004-9615

The entire and exclusive liability and remedy for breach of this Limited Warranty shall be limited to replacement of defective diskette(s) and shall not include or extend to any claim for or right to cover any other damages, including but not limited to, loss of profit, data, or use of the software, or special, incidental, or consequential damages or other similar claims, even if McGraw-Hill has been specifically advised as to the possibility of such damages. In no event will McGraw-Hill's liability for any damages to you or any other person ever exceed the lower of suggested list price or actual price paid for the license to use the Product, regardless of any form of the claim.

THE McGRAW-HILL COMPANIES, INC. SPECIFICALLY DISCLAIMS ALL OTHER WARRANTIES, EXPRESS OR IMPLIED, INCLUDING BUT NOT LIMITED TO, ANY IMPLIED WARRANT OF MERCHANTABILITY OR FITNESS FOR A PARTICULAR PURPOSE. Specifically, McGraw-Hill makes no representation or warranty that the Product is fit for any particular purpose and any implied warranty of merchantability is limited to the sixty day duration of the Limited Warranty covering the physical diskette(s) only (and not the software or information) and is otherwise expressly and specifically disclaimed.

This Limited Warranty gives you specific legal rights, you may have others which may vary from state to state. Some states do not allow the exclusion of incidental or consequential damages, or the limitation on how long an implied warranty lasts, so some of the above may not apply to you.

This Agreement constitutes the entire agreement between the parties relating to use of the Product. The terms of any purchase order shall have no effect on the terms of this Agreement. Failure of McGraw-Hill to insist at any time on strict compliance with this Agreement shall not constitute a waiver of any rights under this Agreement. This Agreement shall be construed and governed in accordance with the laws of New York. If any provision of this Agreement is held to be contrary to law, that provision will be enforced to the maximum extent permissible and the remaining provisions will remain in force and effect.

About the Author

Marcus Goncalves holds an MS in CIS and has several years experience of internetworking consulting in the IS&T arena. He is a Senior IT Analyst for Automation Research Corporation, advising manufacturers on IT, industry automation and Internetworking security. He has taught several workshops and seminars on IS and Internet security in the U.S. and internationally and has published several books related to the subject including *Firewalls Complete* (McGraw-Hill); *Protecting Your Web Site With Firewalls* (PTR/Prentice Hall); *Internet Privacy Kit* (Que); *Windows NT Server Security* (PTR); *IPv6 Networks* (McGraw-Hill); and others. He is also a regular contributor to magazines such as *BackOffice*, *Developer's* and *WEBster*. He is a member of the Internet Society, the International Computer Security Association (ICSA), the Association for Information Systems (AIS) and the New York Academy of Sciences. Goncalves is the chief editor of the *Journal for Internet Security* (JSec).